Earth Science

God's World, Our Home

A Mastery-Oriented Curriculum

Austin, Texas
2015

Earth Science

God's World, Our Home

A Mastery-Oriented Curriculum

Kevin Nelstead

Austin, Texas
2015

Published by

Novare Science & Math
P. O. Box 92934
Austin, Texas 78709-2934
novarescienceandmath.com

Printed in the United States of America

ISBN: 978-0-9966771-9-6 (This ISBN applies only to this preprint (Part 2) edition of the text.)

Cover design by Julie Kennedy, Digital City Designs, www.DigitalCityDesigns.com.

For the complete catalog of textbooks and resources available from Novare Science and Math, visit novarescienceandmath.com.

PREPRINT NOTICE

This volume is Part 2 of a PREPRINT of the forthcoming title *Earth Science: God's World, Our Home*, by Kevin Nelstead, published by Novare Science & Math. Part 1 of the PREPRINT contains Chapters 1–9. This volume contains Chapters 10–13. Also included are Experiments 5 and 6, which are designed for placement after Chapters 8 and 9, respectively. Please note that content in this volume is preliminary and final editing and review of all the chapters has not been completed.

Acknowledgements

We wish to express our appreciation to the author of this text, Kevin Nelstead, who has contributed such an excellent treatment to our textbook line-up.

We are indebted to those who reviewed this text during its development: Chris Mack, PhD; Steven Mittwede, PhD, and Ron DeHaas, and deeply appreciate their work. Thanks also to our excellent copy editor Emily Cook.

The Author

Kevin Nelstead earned a Bachelor of Science degree in Earth Science from Montana State University, and a Master of Science degree in Geology from Washington State University. His Master's degree research involved the study of Quaternary (ice age) volcanic ash deposits in the Channeled Scablands and Palouse Hills area of eastern Washington.

Kevin has seventeen years of experience working as a Senior Cartographer, Geospatial Analyst, and Natural Resources Specialist for the United States federal government. Most of this work has involved creation of topographic maps and interpretation of aerial and satellite imagery. His maps and atlases have placed as high as second in national and statewide professional map competitions.

After earning his teaching certificate in Chemistry from the University of Missouri-St. Louis, Kevin taught science in three different Christian schools: a Roman Catholic high school, a classical Christian school, and an international Christian school. Kevin also regularly teaches Geographic Information Systems courses through his work, and taught as an instructor for one year at Montana State University-Billings.

Kevin has been married to his high school sweetheart, Shirley, for over thirty years, and has four adult children. Kevin and Shirley were missionaries in Romania for over five years, serving with ReachGlobal, which is the international ministry of the Evangelical Free Church of America. They are now members of a Presbyterian Church in America church in Montana.

When he has spare time, Kevin writes about Christianity, science, and the environment at GeoChristian.com.

Reviewers

Steven K. Mittwede	Earth Science & Theology Teacher, Providence Classical School, Spring, Texas PhD, University of South Carolina, Geology MS, University of South Carolina, Geology BS, The College of William and Mary in Virginia, Geology EdS candidate, Columbia International University, Educational Leadership MTh, University of Glamorgan/Wales Evangelical School of Theology, Modern Evangelical Theology MA, Columbia International University, Intercultural Studies
Ronald J. DeHaas	Consulting Petroleum Geologist, Michigan Basin and CEO, Covenant Eyes, Inc. PhD Candidate, Geology, University of Michigan MS, Geology, The Ohio State University BS, Geology, The Ohio State University
Chris Mack	Adjunct Faculty, University of Texas at Austin PhD, University of Texas at Austin, Chemical Engineering MS, Electrical Engineering, University of Maryland BS degrees in Physics, Electrical Engineering, Chemistry, Chemical Engineering, Rose-Hulman Institute of Technology

Earth Science

God's World, Our Home

Contents

PREPRINT NOTICE

This volume is Part 2 of a PREPRINT of the forthcoming title *Earth Science: God's World, Our Home*, by Kevin Nelstead, published by Novare Science & Math. Part 1 of the PREPRINT contains Chapters 1–9. This volume contains Chapters 10–13. Also included are Experiments 5 and 6, which are designed for placement after Chapters 8 and 9, respectively. Please note that content in this volume is preliminary and final editing and review of all the chapters has not been completed.

Experimental Investigation 5: Modeling Weathering

Overview

The goal of this experiment is to observe and measure the effects of mechanical and chemical weathering on a substance. Plants produce many different types of sugars. Table sugar is a sugar called sucrose, which has the chemical formula $C_{12}H_{22}O_{11}$. In this experiment, you will use sugar cubes as a substitute for minerals.

As you have learned, one way that mechanical weathering occurs is when individual mineral grains are transported by water, wind, gravity, or ice. As mineral grains are transported, they bump into each other and are abraded. In this experiment, you will model transport of mineral grains by shaking sugar cubes in a container and observing what happens to them. Because a large number of small mineral grains have a larger surface area than a small number of large mineral grains, chemical weathering occurs more rapidly when grain size is smaller. This is illustrated in Figure 8.6. You will model this by comparing the dissolution[1] rate of whole sugar cubes and crushed sugar cubes.[2] You will also model the effect of temperature on the rate of weathering.

Basic Materials List

- sugar cubes
- plastic shaker bottle with lid
- weighing scale
- weighing paper
- two 250 mL beakers
- warm and cold water
- stopwatch

Safety Precautions

- Follow all safety instructions given by your teacher, such as wearing safety goggles and other protective equipment.

---

1 The term *dissolution* refers to the process of a substance dissolving into another substance.

2 Note that the dissolution of table sugar in water is not actually chemical weathering, as the dissolved substance is still sugar, $C_{12}H_{22}O_{11}$. In chemical weathering, one substance is changed into different substances.

- Do not eat the sugar cubes. Once the sugar cubes are in your laboratory space, they have come into contact with other chemicals used in previous experiments.

Part 1—Mechanical Weathering

In order to model mechanical weathering, you will place five sugar cubes in a sealed bottle and shake them, observing and measuring how the sugar cubes change over time. Throughout this experiment, always handle the sugar cubes with care, being careful to not break off small sugar grains when handling them.

1. Create a data table in your lab journal. Make sure that the boxes in the second column have sufficient space in which to sketch a sugar cube.

Shaking Trial	Drawing of Sugar Cube	Mass of Sugar Cube + Weighing Paper	Mass of sugar cubes (g)
0 shakes			
After 20 shakes			
After 40 shakes			
After 60 shakes			
After 80 shakes			
After 100 shakes			

2. In the "0 shakes" row of your data table, make a sketch of what a typical sugar cube looks like prior to shaking.
3. Determine the mass of your piece of weighing paper by itself and record the value in your lab journal.

4. Place five sugar cubes on your piece of weighing paper and determine the mass. Record the mass of the sugar cubes plus weighing paper in the data table. Subtract the mass of the weighing paper and record the mass of the sugar cubes in the final column.
5. Carefully place the sugar cubes in the shaking bottle and close the lid.
6. Shake your sugar cubes 20 times. Pour the sugar cubes onto your weighing paper. If any sugar crumbs pour out onto the weighing paper, pour them back into the shaking bottle. Determine the mass of the weighing paper plus sugar cubes and record the value in the data table. Again, subtract the mass of the weighing paper and record the mass of the sugar cubes in the final column.
7. Repeat steps 5 and 6 until the data table is complete. For the sake of consistency, try to shake the bottle with the same amount of energy each step. For each step, leave the sugar crumbs in the shaking bottle along with the sugar cubes.
8. Create a graph showing how the mass varies with the number of shakes.

Questions

1. What changes did you notice in the sugar cubes?
2. What caused the sugar cubes to change?
3. What was the end product of the shaking? Predict what would happen if you were to shake the bottle another 100 times.
4. How does this part of the experiment relate to the weathering of rocks and minerals in nature?

Part 2—Chemical Weathering

There are two sections to this part of the experiment. In the first section, you compare the rate of dissolution for a whole sugar cube to that of a crushed sugar cube. In the second section, you compare the rate of dissolution for a sugar cube in cold and warm water.

1. Create a data table in your lab notebook:

Form	Time to dissolve (sec)
Whole sugar cube	
Crushed sugar cube	

2. Place two sugar cubes on a piece of weighing paper.
3. Carefully crush one of the sugar cubes, being careful to not lose any of the sugar granules. You can use the bottom of a beaker to crush the sugar cube.
4. Pour enough cold water into each empty beaker to cover a sugar cube. Pour the same amount of water into each beaker.

5. Place the sugar cube in one beaker and use a stopwatch to measure the amount of time it takes to dissolve. Record the time in the data table.
6. Pour the crushed sugar from the other sugar cube into the other beaker and use a stopwatch to measure the amount of time it takes to dissolve. Record the time on the data table.
7. Create another data table in your lab notebook:

Temperature	Time to dissolve (sec)
Sugar cube in cold water	
Sugar cube in warm water	

8. Pour cold water into one beaker, and warm water into another beaker. Pour the same amount of water into each beaker.
9. Place a sugar cube into the cold-water beaker and use a stopwatch to measure the amount of time it takes to dissolve. Record the time in the data table.
10. Place a sugar cube into the warm-water beaker and use a stopwatch to measure the amount of time it takes to dissolve. Record the time in the data table.

Questions

1. Which dissolved more quickly, a whole sugar cube, or a crushed sugar cube? Why?
2. Which sugar cube dissolved more quickly, the one in the cold water or the one in the warm water? Why?
3. How does this part of the experiment relate to the weathering of rocks and minerals in nature?
4. Table sugar is not considered to be a mineral. Why not?

Experimental Investigation 6: The Stream Table

Before the development of computer modeling, engineers built scale models of streams in order to investigate the effects of stream processes such as erosion, deposition, and flooding. This photograph shows a model used in the 1930s to study flooding in central Pennsylvania. Physical models, such as the stream table, are still useful for learning how these processes work.

Overview

The goal of this experiment is to model stream erosion and deposition using a stream table. In this investigation, you will model stream flow using a stream table, which is a flat, open-topped tray or box containing sand and other sediments through which a small stream is allowed to flow. Stream tables are an excellent tool for visualizing stream processes such as deposition, erosion, stream channel changes, and growth of deltas.

Basic Materials List

- stream table
- sand
- water
- table knife
- thin board that can be used as a dam

Safety Precautions

- Follow all safety instructions given by your teacher, such as wearing safety goggles and other protective equipment.
- Water and sand should stay inside the table. Do not splash the water.
- You should wash your hands after using the stream table.
- If your stream table has an electric water pump, it is essential that you keep sand and water away from the electrical outlet.

Procedural Notes

One of the most important things to remember as you go through the steps of this investigation is the importance of consistency. Try to have a steady stream flow throughout each part of the experiment. Try to use the same rate of stream flow in each step. You will be asked to draw sketches, or maps, of the stream table several times throughout the investigation. Make sure each sketch has a title, such as "Meandering Stream" or "Meandering Stream after 10 Minutes." It will be helpful to read through the questions at the end of this experiment now instead of waiting until you are done.

Stream Table Setup

1. Fill the upstream end of the stream table with sand. Raise the sand-filled end and gently shake the stream table until the sand covers about the top two-thirds of the stream table. It is important to leave an empty space at the foot of the stream table as an area for water to pool and for a delta to form.
2. Raise the upstream end of the stream table about 3–5 centimeters.

First Stream

3. Begin slowly and steadily pouring water at the upstream end. Observe the small stream that forms in the sand, paying attention to places where erosion and deposition are occurring.
4. In your lab notebook, draw a sketch of the stream table as viewed from above, showing the stream channel. Label areas of erosion and deposition.
5. Allow the stream to flow for a few minutes and observe how the stream changes over time.

Meandering Stream

6. Turn off the stream flow. With your fingers, change the stream to be a winding, meandering channel. It is better to make larger meander loops rather than small ones. If you create a meander loop with a narrow neck, you will be able to observe erosion through the neck to create an abandoned meander and perhaps an oxbow lake.
7. Start a gentle stream flow again. In your lab journal, draw a sketch of the stream table as viewed from above, showing the stream channel. Label areas of erosion and deposition. Also label the cut bank and point bar.
8. Observe the stream channel for a few minutes. You should be able to see your meander migrate as erosion occurs on cut banks and deposition occurs on point bars. Draw another sketch showing how the channel has changed over time.

Stream Delta

9. You should have a delta or alluvial fan formed at the mouth of your stream. Draw a sketch of this landform, showing the details of the channels that form.
10. Stop the stream flow and carefully remove water from the foot of the stream table. Slice into the delta with a table knife. If layering is visible, make a sketch of the layers in your lab journal.

Changing the Stream Gradient

11. Raise the upstream end of the stream table by 2–3 centimeters.
12. Start a gentle stream flow again, similar to the previous flow. Observe how the change of stream gradient has affected the rate of erosion. If there is no observable change to the rate of erosion, raise the upstream end of the stream table by a few more centimeters.

Placing a Dam

13. Use a board or other object to place a dam across the stream channel.
14. In your lab notebook, draw a sketch of the stream table, with labels for areas of erosion and deposition.

Additional Activities

Depending on the amount of time you have, there are many other things you can do with a stream table, such as modeling floods, erosion control, effects of dam removal, effects of changing sea level, beach erosion, and much more.

Questions

1. Your first stream was probably fairly straight. How did its channel change over time?
2. Describe how sand grains move along the bottom of the stream channel. Do they move by bouncing, rolling, floating, or by some other means? Do individual sand grains move continuously from the top of the stream to the delta? If not, how far do sand grains typically move at a time?
3. Where did erosion occur along your meandering stream? Where did deposition occur?
4. How did your meandering stream change over time?
5. In what ways did increasing the stream gradient by raising the upstream end of the stream table change the rates of erosion and deposition along your stream?
6. The building of a dam changes a stream both upstream and downstream from the dam. Describe changes to erosion and deposition that occurred along the length of your stream after you added the dam.

Chapter 10

Landforms

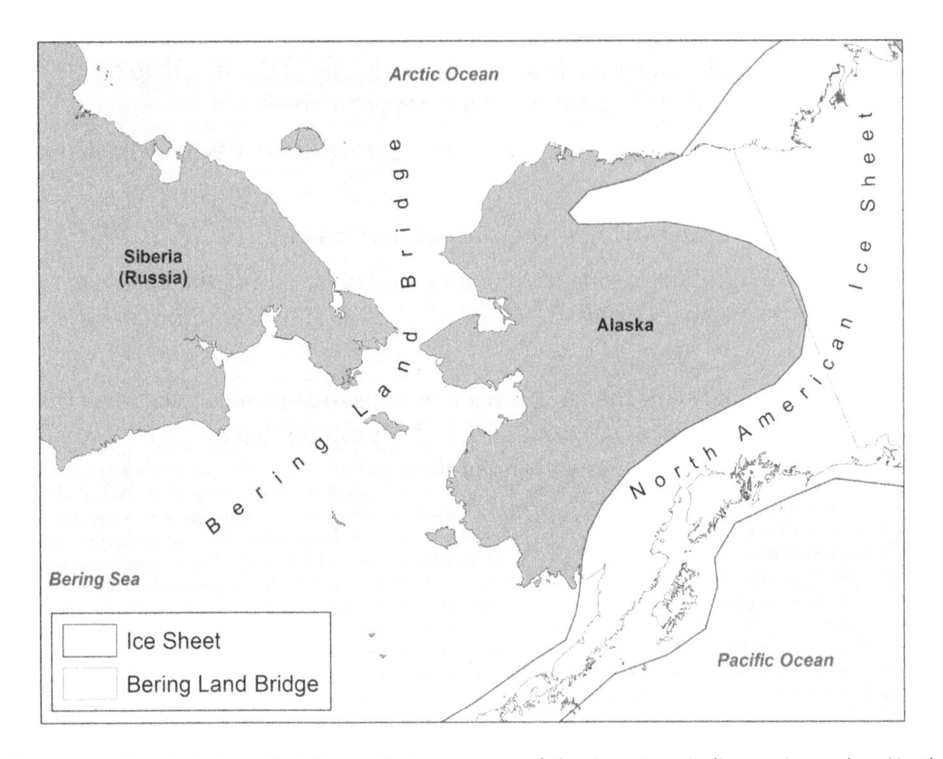

Most archeologists believe that the earliest ancestors of the American Indians migrated to North America sometime between 16,000 and 13,000 years ago. For most of human prehistory—that time before widespread use of writing—people only lived in the Eastern Hemisphere; primarily on the continents of Africa and Eurasia. How did the ancestors of today's Native Americans arrive in the Western Hemisphere? Most scientists who have studied that issue have concluded that they walked from Siberia to Alaska. If you know your geography, you know that Siberia and Alaska are separated by the Bering Strait, and that it is now impossible to walk from Asia to North America. So why do scientists believe it was once possible?

The Bering Strait is shallow. During Earth's most recent Ice Ages—prior to about 11,000 years ago—much of northern North America and Eurasia was covered by vast sheets of ice, in many places more than 1,000 meters (3,000 feet) thick. These ice sheets were formed through the ordinary processes outlined as the hydrologic cycle in Chapter 9—evaporation of seawater and precipitation as snowfall—during a time when Earth's overall climate was several degrees cooler than it is today. Because the precipitation fell as snow that did not melt and return to the oceans, sea level around the globe fell by about 120 meters. Large areas of the continental shelf that were once under water were then exposed as dry land, including what is now known as the Bering Land Bridge, which allowed humans to walk from Asia to North America and a number of animal species to migrate between the continents in both directions. As glaciers melted over the following millennia, humans were able to migrate southward to populate both North and South America.

In this chapter, you will learn about glaciers and other agents that have worked to shape Earth's surface, such as water, gravity, and wind.

Objectives

After studying this chapter and completing the exercises, you should be able to do each of the following tasks, using supporting terms and principles as necessary.

1. Describe the variables that influence the occurrence of mass movement.
2. Explain how the different types of mass movement vary from one another.
3. Describe the role of water in desert landscapes.
4. Compare the transport of sediments by wind and by water.
5. Describe types of wind deposits and explain how each forms.
6. Describe how snow transforms to make glacial ice, and how glacial ice behaves differently than ordinary ice.
7. Explain how glaciers move.
8. Describe types of glacial erosional and depositional landforms, and explain how each forms.
9. State reasons why geologists believe vast ice sheets and alpine glaciers once covered large parts of Earth's surface that are now ice-free.

Vocabulary Terms

You should be able to define or describe each of these terms in a complete sentence or paragraph.

1. abrasion
2. alpine glacier
3. arête
4. barchan
5. cirque
6. creep
7. crevasse
8. debris flow
9. deflation
10. desert pavement
11. drumlin
12. dune
13. ephemeral stream
14. erratic
15. esker
16. fjord
17. glacial striation
18. glacier
19. ground moraine
20. hanging valley
21. horn
22. ice age
23. ice sheet
24. internal flow
25. kettle
26. landform
27. landslide
28. loess
29. longitudinal dune
30. mass movement
31. moraine
32. parabolic dune
33. plucking
34. rock fall
35. saltation
36. slip face
37. slump
38. snowline
39. tarn
40. till
41. transverse dune

10.1 Landforms Caused By Mass Movement

Earth's surface is covered by a wide variety of distinctive features known as *landforms*: hills, mountains, ridges, valleys, cliffs, beaches, sand dunes, and many more. You have already learned about some landforms, such as cinder cones, alluvial fans, floodplains, oxbow lakes, and sinkholes. In this chapter, you will learn about more landforms, each of which is created by forces acting on materials on Earth's surface.

Most of Earth's surface is not horizontal; instead, most landscapes have some degree of slope, ranging from subtle, rolling hills to steep cliffs and towering mountains. On all slopes, the force of gravity constantly pulls material downhill, but downhill movement is resisted by forces that hold grains in place. The result of this battle between gravity and other forces is that material moves downhill on slopes, sometimes extremely slowly, and other times catastrophically.

The downslope movement of material due to the force of gravity is called *mass movement*. On Earth's surface, weathering weakens and breaks apart rocks, but mass movement is the primary means by which the weathered material moves downhill, eventually ending up in streams, where the material can be transported elsewhere. This can be seen by looking at any valley or canyon, such as the Grand Canyon in Figure 8.16. A stream can only erode at the bottom of a canyon; streamflow has no direct effect on rocks and soil on the slopes above the river. Weathered material from the slopes of the canyon moves downhill primarily through mass movement.

Figure 10.1. If sand is too dry, sand grains do not stick to each other to make vertical walls like the ones in this sand castle. If sand is moist, the sand flows as a slurry of sand and water.

10.1.1 Variables That Influence Mass Movement

There are several factors that influence what types of mass movement will occur on a slope and the rate at which material will move down a slope. One of the most important of these factors is the amount of water in loose material on a slope. The presence of some water helps particles to stick together,

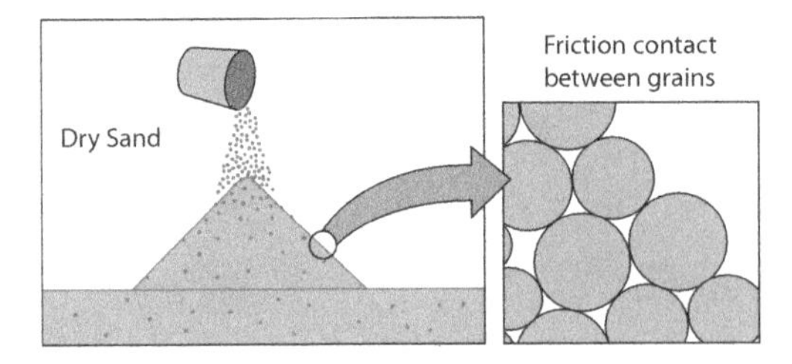

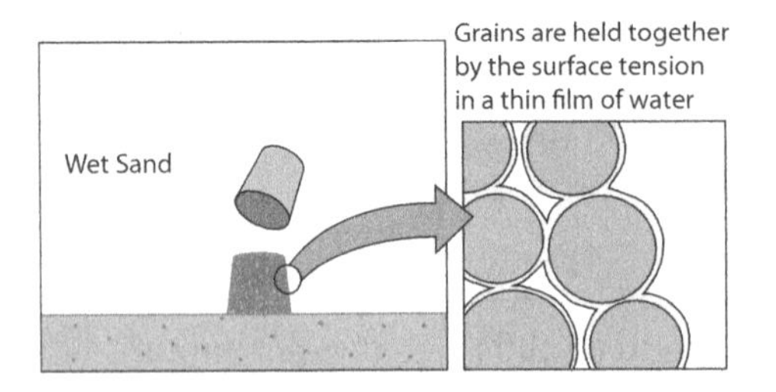

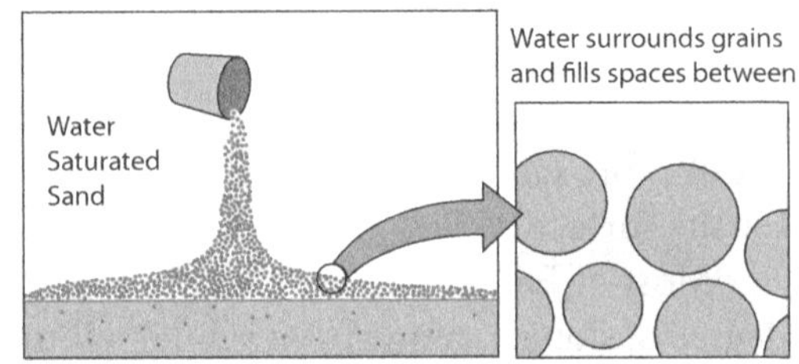

Figure 10.2. The amount of water in pore spaces in sand controls how sand grains stack. Dry sand will make a cone with an angle around 35°, moist sand can be piled with an almost vertical wall, and water-saturated sand will flow as a fluid.

reducing the likelihood of movement, but having too much water in soil makes it more likely that it will move downhill. Figure 10.1 shows a very finely crafted sand castle. If you have ever built a sand castle, you know that the grains in dry sand do not stick to each other, so dry sand cannot be used to make a vertical wall. Dry sand will flow and make piles with a slope of about 35°, as illustrated in Figure 10.2 Adding a little bit of water to the sand makes the sand grains stick to each other, allowing you to make castle turrets and walls. The stickiness is caused by thin films of water between the sand grains. But if more water is added, the water will act as a lubricant rather than helping the grains stick to each other and the sand will flow easily. Water acts the same way in unconsolidated material on a hillside. A certain amount of water helps to hold material in place, but a large amount of water, such as you might find after a heavy rain, often causes material to move downslope.

Figure 10.3. A small alluvial fan formed after vegetation was removed from a landscape by a wildfire. The slopes in this area are much more susceptible to mass movements and erosion until grass and other vegetation grow back.

Another important factor affecting mass movement is the presence of vegetation on a surface. Plant roots help to hold soil in place on hillsides. If plants are sparse, such as in arid regions, or if plants are removed through natural means such as fires or by human activities, then the soil on the hillside will be more susceptible to mass movement. Figure 10.3 shows an area where vegetation has been removed by a forest fire.

Mass movement frequently occurs along the banks of rivers. When a stream undercuts its bank, gravity pulls material down into the stream, as shown in Figure 10.4. The same thing is common along coastlines where waves erode the base of cliffs or hillsides causing material to slide down toward the shore. Mass movement can also be caused or

Figure 10.4. The stream in the foreground has undercut its bank, but the soil is being held in place by tree roots. Eventually, the bank will collapse into the stream.

accelerated by earthquakes. In some earthquakes, more damage is done by landslides than by shaking.

10.1.2 Types of Mass Movement

The slowest type of mass movement is *creep*, which is the slow, downslope movement of soil. Creep occurs at rates measured in millimeters or centimeters per year, so it cannot be directly observed. The effects of creep, however, are visible on many hillsides. As illustrated in Figure 10.5, creep causes fence posts and telephone poles to tilt downhill, tree trunks to curve, and retaining walls to break.

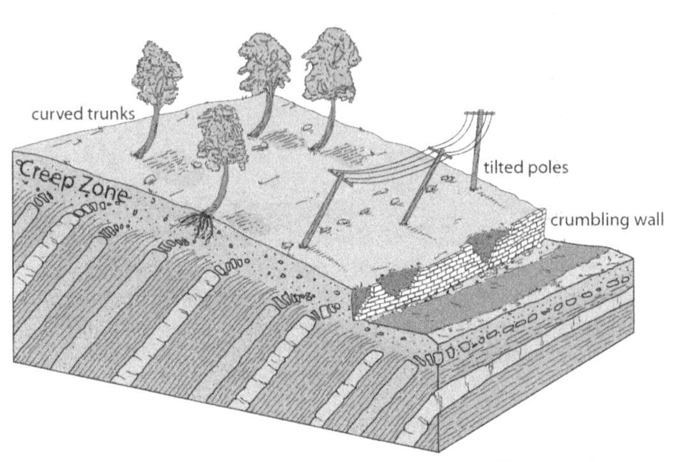

Figure 10.5. Creep occurs at a very slow rate, but its effects can become obvious over a period of many years.

Other types of mass movement, illustrated in

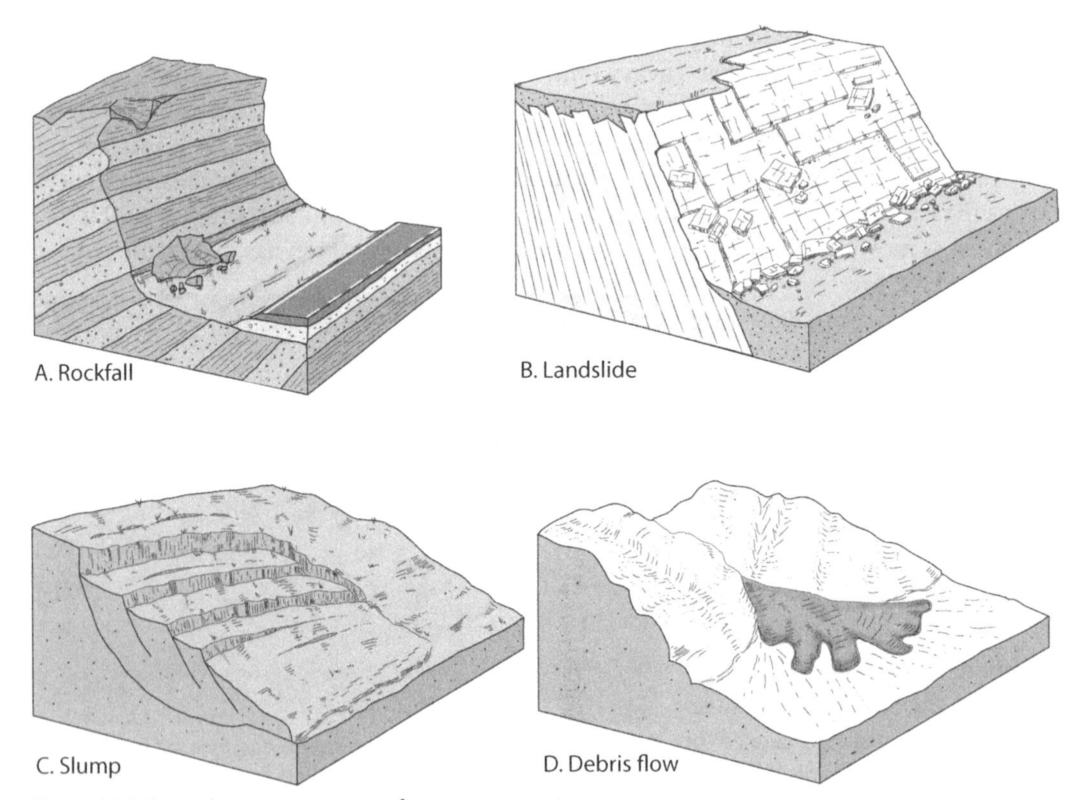

Figure 10.6. Four other common types of mass movement.

Figure 10.6, happen more rapidly. The most rapid type of mass movement is a *rock fall*, in which rocks fall freely through the air for much of their downslope journey. Falls can only happen along cliffs and other very steep slopes, and may involve rocks rolling and bouncing down the slope. Rock falls are a common hazard along roads in mountainous areas, as you can see in Figure 10.7. Often, the rocks accumulate at the base of the slope as a talus slope, shown in Figure 8.2.

Figure 10.7. A rock fall blocked this mountain highway.

A *landslide* is a sudden movement in which loose rock and soil travels down a slope. Landslides are common in hilly and mountainous areas and cause over $2 billion in damages in the United States each year. Figure 10.8 shows a landslide that killed 43 people near Oso, Washington in 2014. Geologists have observed that the hills along the valley where the Oso landslide occurred show scars of large prehis-

Figure 10.8. The Oso Landslide in Washington in 2014 killed 43 people.

Figure 10.9. A slump showing the crescent-shaped upper border that is typical of slumps. Each block of soil has rotated as it moved downslope.

Figure 10.10. A debris flow destroyed this home in California. Note that the mud contains boulders that were carried along as part of the debris flow.

toric landslides, and it would be unwise to allow further construction in these landslide hazard zones.

A *slump* is similar to a slide, but the material moves along a curved surface, as shown in Figure 10.6C. Slumps commonly have a crescent-shaped head where the slump separates from the remaining hillside, and a pile of debris that accumulates at the toe of the slump. Slumps only occur in loose material, such as soil, and the material rotates as it slides along a curved surface. Figure 10.9 shows a slump.

A *debris flow* is a mass movement in which unconsolidated moving material is saturated with water. Some debris flows are quite watery, but many flows are a thick mixture of everything from clay, silt and sand to large boulders and trees that have been washed off of hillsides. If a debris flow is composed primarily of mud and water, it is called a *mudflow*. Debris flows often travel rapidly; usually along the bottoms of valleys. Debris flows are most common in semiarid regions such as much of southern California, and occur during periods of heavy rain or rapid snowmelt. Figure 10.10 shows the deposits of a debris flow near Los Angeles.

A lahar (described in Section 7.2.2) is a variety of debris flow associated with volcanic eruptions when volcanic ash is mixed with water. Rainwater mixing with volcanic ash can cause lahars, but they can also be formed by the rapid melting of snow and ice during an eruption.

Learning Check 10.1

1. What is the force that causes all mass movement, and what factors prevent loose material from sliding downhill?
2. Compare and contrast five types of mass movement.
3. Describe the role of water in causing or preventing mass movement.

10.2 Desert Landforms

Almost one third of Earth's land surface is covered by arid and semi-arid regions, such as deserts, dry grasslands, and shrublands. Because of the scarcity of water, a number of distinctive landforms occur in dry landscapes. Most desert areas are covered by bare rock or gravelly surfaces rather than by sand sheets or sand dunes.

10.2.1 Streams in the Desert

A desert is a place where there is not enough precipitation to support abundant plant life. In general, deserts are places where the amount of evaporation is greater than the amount of precipitation, though no deserts are completely rainless. When rain does fall in deserts, streams rapidly fill up with water, and in many places long-dormant plants flourish for a short period of time.

A stream that only flows occasionally is called an *ephemeral stream*. Some ephemeral streams have water in them for a few days or weeks during the wettest part of the year, while ephemeral streams in some regions may not carry water for years at a time. These ephemeral streams are sometimes called "washes" or "arroyos" in the southwestern United States. In North Africa and the Arabian Peninsula, an ephemeral stream is referred to as a "wadi," a term that occurs in some translations of the Old Testament. Figure 10.11 shows an ephemeral stream bed.

Figure 10.11. A wadi, or ephemeral stream, in the Negev Desert of Israel.

10.2.2 Wind Erosion and Transport

Places with barren ground, such as deserts, beaches, or newly deposited sediments, are more vulnerable to wind erosion than others. Plants help to protect against wind erosion, so places with trees or grass experience little wind erosion. Soil moisture also reduces wind erosion because water helps grains stick together.

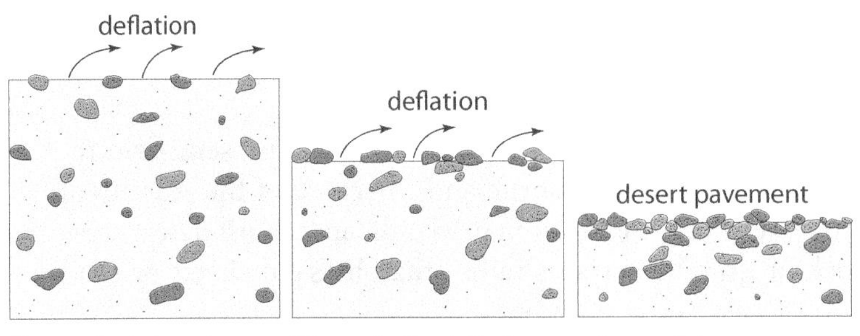

Figure 10.12. As a desert soil is eroded by deflation, smaller particles are removed, leaving a desert pavement behind. Once desert pavement forms, no further erosion by deflation can occur.

One way wind erodes Earth's surface is by *deflation*, the lifting and removal of loose material by wind. During deflation, wind removes the top layer of fine soil particles. In some places, the material left behind, mostly gravel, forms a surface of tightly-packed rocks called *desert pavement*, common in desert areas. The creation of desert pavement by deflation is illustrated in Figure 10.12, though not all desert pavements form by the same process. Desert pavement protects the underlying material from further erosion. Figure 10.13 shows a desert pavement.

Figure 10.13. Desert pavement is a gravelly surface common in deserts. The ground surface has little or no sand, silt, or clay.

Deflation is a significant erosional agent in some areas. The dust storms that occurred during the Dust Bowl, discussed on page 200, removed millions of tons of topsoil from the Great Plains in the 1930s. When deflation removes all topsoil from an area, the wind sometimes creates shallow depressions called *deflation hollows* (or *blowouts*) on the landscape, as shown in Figure 10.14. Deflation hollows may be further eroded by wind, enlarging to a width of several kilometers.

Figure 10.14. Deflation hollows, such as this one in Texas, are depressions formed by wind erosion.

Wind also erodes by *abrasion*, which is the grinding and erosion of rock caused by the impact of rock particles carried by wind, water, or ice. Figure 10.15 shows a rock on a desert surface that has been abraded by wind-blown sand.

Figure 10.15. This rock has been abraded by wind-blown sand.

Wind moves different types of sediment in different ways. Dust—that is, silt and clay—is picked up by wind, carried to high altitudes, and transported long distances. This is illustrated by Figure 8.13, a satellite image showing vast clouds of dust being carried from the Sahara Desert across the Atlantic Ocean. Figure 10.16 shows a dust storm from Earth's surface.

In a dust storm, silt and clay are carried by suspension, just as fine-grained sediments are carried by suspension in running water. Sand grains are also carried by wind. However, wind cannot hold sand grains up in the air for long, so sand grains are transported by bouncing along the ground. This process is called *saltation*, illustrated in Figure 10.17. Sand grains begin to move when the wind blows fast enough to roll some of the sand grains along the surface, a process called "surface creep." These moving sand grains strike against other grains, causing one or both grains to jump into the air. Once in the air, the grains are carried forward by the wind, but immediately start falling toward the ground because of gravity. When these saltating sand grains strike the ground, they either bounce back into the air, knock other grains up into the air, or both.

Figure 10.16. Silt and clay, together known as dust, can be transported by suspension high above Earth's surface, as in this dust storm in Texas.

Even when the wind is blowing strongly, sand grains seldom bounce more than a meter into the air, so saltation only occurs close to the ground. Particles moving by saltation and surface creep represent the majority of wind-blown sediment transport.

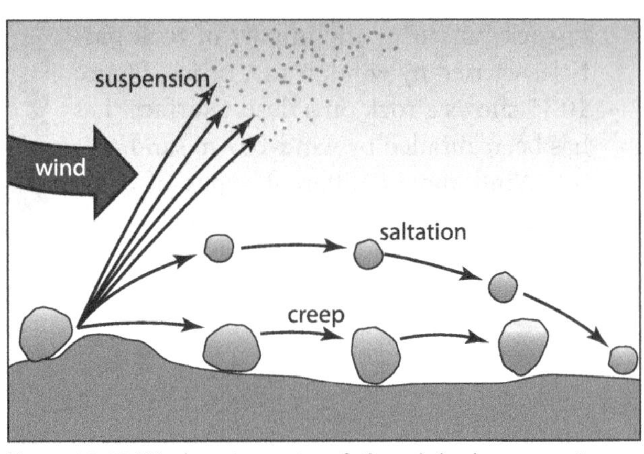

Figure 10.17. Wind carries grains of silt and clay by suspension. Sand grains mostly transport by saltation, a process of jumping or bouncing along the surface.

10.2.3 Wind Deposits

The most familiar landform created by wind is a *dune*—a hill or ridge of wind-deposited sand. Dunes are common in deserts as well as along the shores of lakes, seas, and oceans. A dune begins to form when some sort of barrier, such as a rock or a bush, slows wind down, causing sand to begin to accumulate downwind of the barrier. The pile of sand soon becomes a barrier itself, and the dune grows larger and larger.

Many dunes have a gentle slope facing into the wind, and a steeper slope, called the *slip face*, on the downwind side, as illustrated in Figure 10.18. As wind blows, sand moves up the gentle slope by saltation. When the sand reaches the crest of the dune, wind speed drops, and sand particles are deposited at the top of the slip face. As sand accumulates on the slip face, the slip face becomes steeper and sand begins to slide down the slope. Over time, the gentle upwind side of the dune erodes and particles are transferred to the slip face, causing the dune slowly to move across the landscape in the direction the wind is blowing.

Dunes are classified by their shapes, as illustrated in Figure 10.19. *Barchan*[1] dunes are solitary, crescent-shaped dunes that move across flat areas where there is little vegetation. Barchans form where sand supply is limited and where the wind blows from roughly the same direction year round. The "horns" of a barchan point downwind. Figure 10.20 shows barchans on the surface of Mars. *Transverse dunes* form long, flattened ridges that are perpendicular to the wind direction. Transverse dunes form in areas with an abundant supply of sand and strong winds. *Longitudinal dunes* also form in long, parallel ridges, but longitudinal dunes run parallel to the direction of wind. Longitudinal dunes, like barchans, form where

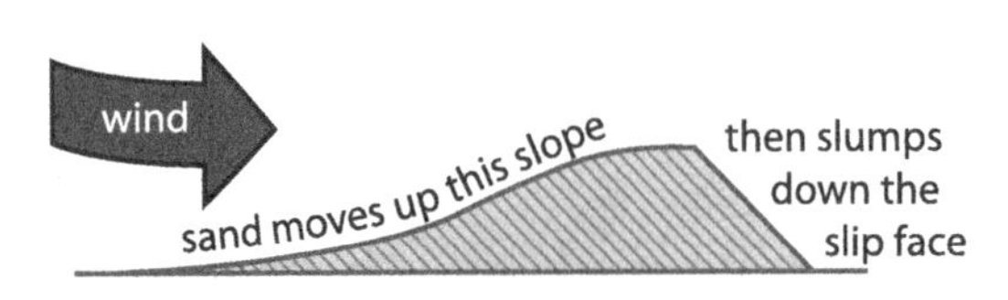

Figure 10.18. Sand moves up the gentle slope of a dune by saltation, and then cascades down the slip face. The result is that the dune migrates in the direction of wind movement.

1 Pronounced BAR-kan.

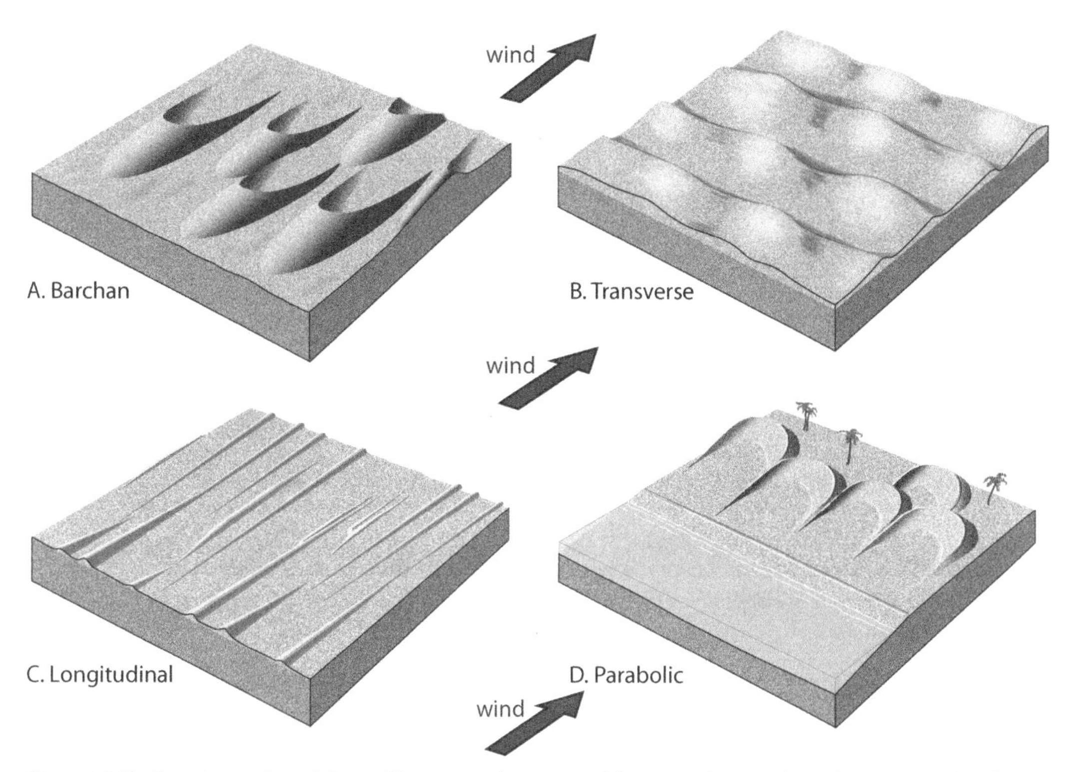

Figure 10.19. Four types of sand dunes. There are other types of dunes, and many dunes have properties that combine two dune types.

sand is available in limited amounts. *Parabolic dunes* are U-shaped, somewhat like barchans, but the "horns" point upwind instead of downwind. Parabolic dunes only form where there is some vegetation, which helps to anchor the dune in place. They are most common in humid areas along coastlines, and are not limited only to deserts.

Some parts of Earth's surface are covered by thick layers of windblown silt deposited as a result of thousands of years of dust storms. This wind-deposited silt is known as *loess*[2] and is usually yellow or buff in color. Thick loess deposits exist on the eastern edges of the Mississippi and Missouri River valleys, as well as in eastern Washington, as shown in Figure 10.21. During the ice ages, melting glaciers deposited silt along rivers in these areas. The silt was later transported by wind to form loess hills. There are also thick deposits of loess in northern China, formed by silt blown out of the Gobi Desert.

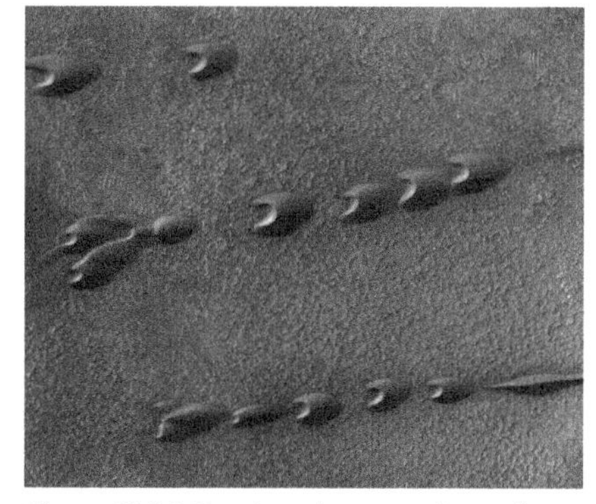

Figure 10.20. Barchan dunes on the surface of Mars. The "horns" of the dunes point to the left, indicating that wind usually blows from the right to the left in this area.

2 *Loess* may be pronounced in a variety of ways, including *luss*.

Figure 10.21. Thick, fertile deposits of loess deposited by wind during the ice ages cover parts of eastern Washington.

Erosion of Chinese loess is what makes the Yellow River and Yellow Sea yellow. Loess deposits are highly fertile and provide excellent soil for crop growth.

Learning Check 10.2

1. Describe factors that make it more likely that an area will experience wind erosion.
2. Describe how wind transports sediments of various sizes and compare that to sediment transportation by water.
3. Describe how sand dunes move.
4. Sketch four types of sand dunes, with labels for wind direction, and brief text describing what makes each type distinct.

10.3 Glaciers

As you learned in Chapter 9, there is far more water stored on Earth's surface in the forms of snow and ice than there is stored as liquid water in streams, lakes, and groundwater. A *glacier*, such as the glacier shown in Figure 10.22, is a large, moving mass of ice on land. A key part of that definition is the word *moving*. The ice in glaciers constantly moves downhill at a very slow rate, and thus glaciers are sometimes referred to as "rivers of ice." The speed at which most glaciers move is just a few millimeters per day. Note that an avalanche is not a glacier because an avalanche is made primarily of snow, not ice, and because an avalanche is a sudden movement that has a distinct beginning and end, while glaciers move gradually and continuously.

Figure 10.22. A large glacier in the Swiss Alps. The ice forms at higher elevations and flows downhill until it melts.

Glaciers form in areas where the temperature is cold enough year-round so that not all snow that falls during cold seasons melts during the summer. These cold places are found in polar regions and in mountain ranges throughout the world. Most of Antarctica and Greenland are covered by thick glacial ice, and glaciers are found in mountain ranges such as the Alps, Himalayas, and Andes. In the United States, large glaciers are found in the mountain ranges of Alaska, smaller glaciers in the mountain ranges of the Pacific Northwest, and even smaller glaciers in the higher parts of the Rocky Mountains. Glaciers also exist in tropical latitudes in locations where mountains are high enough, such as in the Andes in South America, as well as in isolated high mountain ranges in Africa and New Guinea.

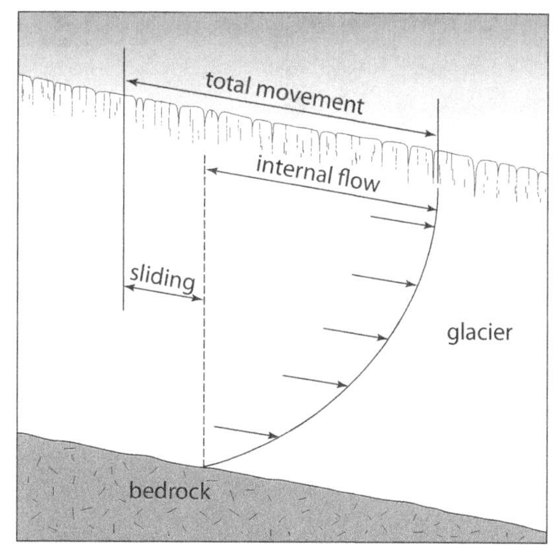

Figure 10.23. A vertical slice through a glacier illustrates ice movement. Glaciers move partly by sliding along their base, and partly by internal flow. Due to friction along the contact between rock and ice, internal flow is slowest along the base and sides of the glacier, and fastest near the glacier's surface.

10.3.1 Glacial Ice

Snow regularly falls on the land in many parts of the world, but in most places that snow melts rather quickly. At high elevations and close to the poles, however, there are places where it is cold enough, even in summer, that not all snow that falls in the winter melts. In these locations, the

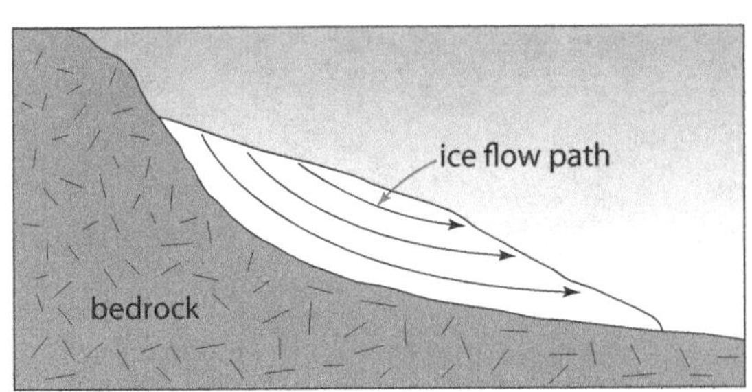

Figure 10.24. Paths taken by ice within a glacier as the ice flows downslope.

snow accumulates to greater depths with each passing year. The elevation above which not all snow melts is called the *snowline*. This elevation varies from above 5,000 meters (over 16,000 feet) near the equator, to near sea level in coastal polar areas.

As old snow becomes more deeply buried by newer snow, it compacts and recrystallizes into tightly packed grains of ice, much like when wet snow is compressed tightly into a snowball. At a burial depth of about 50 meters, the ice is converted into a solid mass of interlocking ice crystals known as "glacial ice." Glacial ice differs from the ice you would get from an ice cube tray in a freezer in that it is under great pressure from the weight of the overlying ice. Unlike the ice cubes you are familiar with, glacial ice is capable of *internal flow* because its microscopic structure is much more able to deform without breaking. With internal flow, parts of the ice are able to move more rapidly than others. This is illustrated in Figure 3.6, which shows how stakes in the middle of a glacier move downhill more rapidly than stakes near the edges of a glacier.

Glacial ice is acted on by the force of gravity and on a slope the glacier will start moving downhill. Glaciers move by a combination of internal flow and sliding along the bottom of the glacier. This sliding occurs because most glaciers have a very thin layer of liquid water at their base, which lubricates the contact between the glacier and bedrock. Glacial movement is illustrated in Figure 10.23.

Figure 10.25. Crevasses are deep fractures in the upper, brittle portion of glaciers. (Notice the house on the brown hill in the background.)

Snow and ice from above the snowline flow down to lower elevations where it is warmer, causing ice loss by melting to be greater than ice gain by snowfall. The farther the glacier flows, the more it melts. Eventually, there is so much melting that the glacier advances no farther down the slope. (A glacier "in retreat" is still advancing downhill, but it is melting faster than its advance.) The paths that ice takes inside a glacier as it flows downhill are illustrated in figure 10.24.

Deep within a glacier, ice flows by plastic deformation. In contrast, ice near the surface of a glacier is brittle and lo-

Figure 10.26. Glacial striations are scratches formed when gravel carried in a glacier abrades the bedrock beneath the glacier.

cally breaks into deep fractures called *crevasses*. These crevasses, such as the one in shown Figure 10.25, can be as much as 50 meters deep, and are a serious hazard for people walking on glaciers, especially when fresh snow drifts over the top of the crevasse.

Glaciers have a tremendous capability of eroding and reshaping Earth's surface. Glaciers erode the land in two ways—*plucking* and *abrasion*. Plucking occurs when water gets into cracks in rocks beneath the glacier. When this water freezes, it expands, enabling the glacier to pull the rock out of place and up into the ice. Abrasion is the process in which this ice-borne material acts like sandpaper: grinding, scratching, and polishing rocks as the ice moves. Because of these two processes, ice at the bottom of a glacier contains a large amount of debris, ranging from boulders down to microscopic silt particles. Ice by itself cannot abrade rocks; abrasion is done by rock and mineral grains within the glacier. Larger rock fragments in glacial ice create long scratches on Earth's surface called *glacial striations*, such as the ones shown in Figure 10.26. The presence of striations is evidence of past glaciation in an area, with the striations providing a record of the direction of ice flow.

10.3.2 Alpine Glaciers

There are two main types of glaciers—*alpine glaciers* (also called *valley glaciers*) and *ice sheets*. An alpine glacier forms in mountainous areas where ice that accumulates at higher elevations flows down through long, narrow valleys. Many

Figure 10.27. Alpine glaciers carve a relatively straight U-shaped valley in bedrock, such as this valley in Rocky Mountain National Park in Colorado.

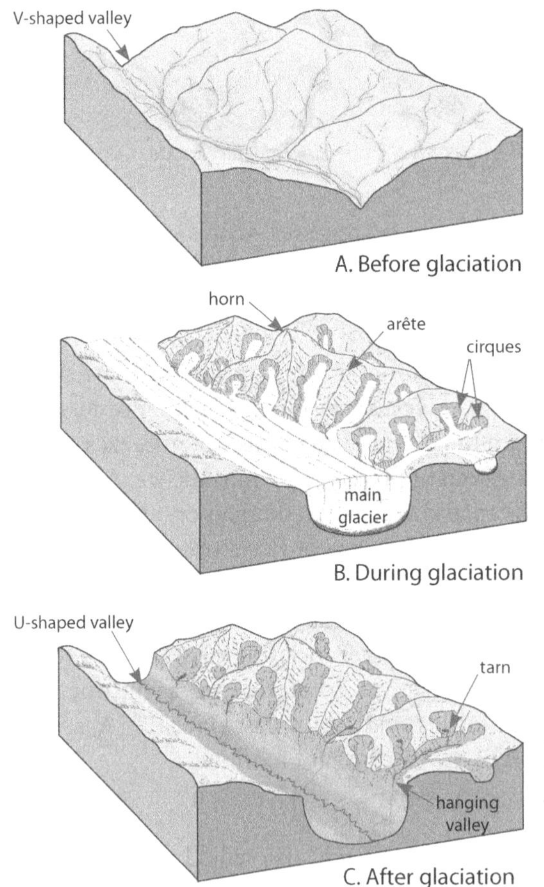

Figure 10.28. Erosional landforms formed by alpine glaciers.

mountain ranges have been sculpted by glacial erosion in the past, and this has created distinct landforms.

Mountain ranges that have never been glaciated often have rounded peaks and ridges, and narrow, V-shaped stream valleys. The erosional work of glaciers creates sharp peaks and ridges, and both widens and straightens valleys, resulting in distinctive U-shaped valleys, such as the one in Figure 10.27.

Figure 10.28 illustrates typical erosional landforms created by alpine glaciers. As a glacier moves through a valley, rocks break off of the valley walls, causing the walls to become steeper. Glacial U-shaped valleys are formed as glaciers pluck and abrade rocks along their sides. Just as large streams have a number of tributary streams, large glaciers often have a number of smaller tributary glaciers. When all of these glaciers melt, the U-shaped valley created by the large glacier will be larger than the valley created by smaller tributary glaciers, leading to the existence of *hanging valleys* that enter the larger valley at a considerably higher elevation. If a valley formed by glaciers near a coastline becomes filled with seawater, a long, narrow inlet of the sea called a *fjord*[3] is formed, such as the one in shown Figure 10.29.

Figure 10.29. A fjord is a U-shaped glacial valley that is filled in with water, usually seawater. This fjord is one of many in Norway.

At the head of the glacier, a bowl-shaped depression is formed, called a *cirque*. Glaciers deeply erode the base of the cirque creating a depression in the bowl. After a glacier melts, this depression will fill with water and the resulting lake is called a *tarn*. Figure 10.30 shows a cirque with a tarn. Sharp ridges separating glacial valleys are called *arêtes*. A pyramid-like peak formed where multiple arêtes meet is called a *horn*. The Matterhorn in the Alps is a classic example of a horn, but horns are actually abundant in many mountain ranges. Figure 10.31 shows hanging valleys, arêtes, and a horn.

Not all glacial landforms are created by erosion. When glaciers melt, they leave behind a tremendous amount of material that was carried in the ice. The general word for sediments deposited directly by glaciers is *till*. Sediments deposited by water are often layered and well-sorted, meaning that the sand is deposited together in layers, silt in different layers, and gravel in still different layers. Till, on the other hand, is typically poorly layered and poorly sorted, meaning that silt, sand, and gravel are all mixed together. Till can form a thick blanket covering the ground, or it can be piled at ridges called *moraines*, which form at the edges of glaciers. A moraine marking the farthest distance that a glacier advanced is called a "terminal" moraine. Glaciers can push material for a short distance,

Figure 10.30. A cirque with a tarn in the Cascade Range of Washington.

3 Pronounced *fyord*.

Figure 10.31. The mountains in Glacier National Park, Montana, have a wide variety of erosional glacial landforms. The sharp ridges on the left are arêtes. The arêtes are separated by U-shaped hanging valleys. The peak in the distance on the right is a horn.

but most of the material deposited in a terminal moraine gets deposited as ice melts at the end of a glacier. Figure 10.32 illustrates three types of moraines associated with alpine glaciers.

10.3.3 Ice Sheets

The longest alpine glaciers in the world today are less than 100 kilometers long, and most alpine glaciers are just a few kilometers long. The second type of glacier is an *ice sheet*,[4] which is a very large, thick mass of glacial ice that spreads out in all directions. A good way to visualize how an ice sheet spreads is by picturing pancake batter spreading in all directions as it is being poured onto a skillet. There are two large ice sheets on Earth today, the Antarctic Ice Sheet and the Greenland Ice Sheet, shown in Figure 10.33, each of which covers millions of square kilometers and is thousands of meters thick. Ice sheets are thickest at the center where there is the greatest accumulation of ice, and thinnest at the edges. At the edges, they either melt on land or flow into the ocean where the ice breaks apart to form icebergs. Earth's ice sheets store millions of cubic kilometers of ice, water that otherwise would be in Earth's oceans. If this ice were completely to melt (which is unlikely),[5] sea level would rise by over 60 meters around the globe. As another illustration of

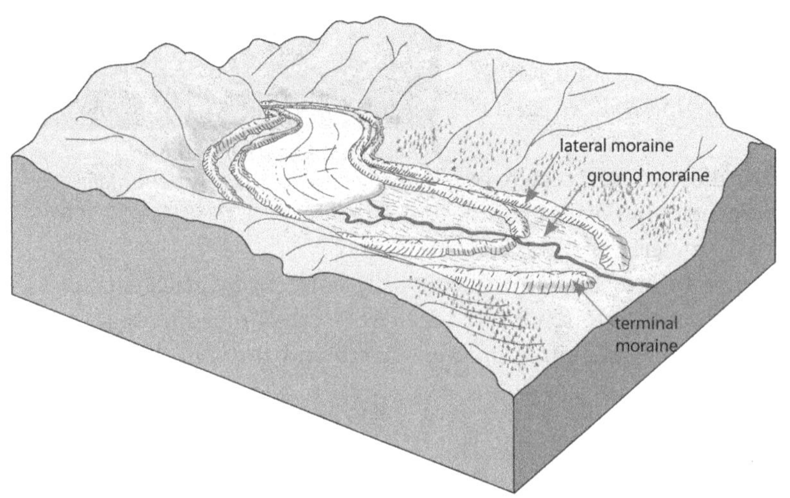

Figure 10.32. Moraines: terminal, lateral, and ground.

4 Ice sheets are sometimes referred to as *continental glaciers*.

5 Even though Earth is warming and the ice sheets are melting, as nearly all climate scientists agree, it is unlikely that the Greenland and Antarctic ice sheets will completely melt unless the amount of warming significantly exceeds current predictions.

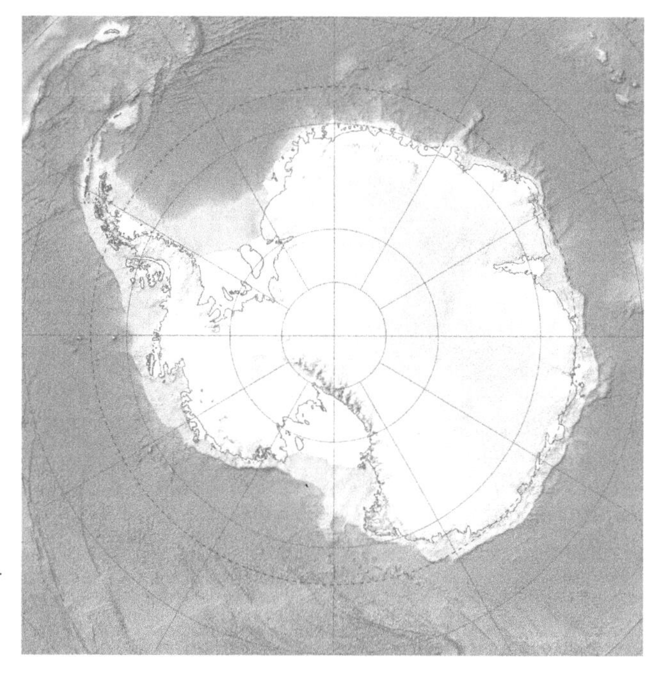

Figure 10.33. Both Antarctica and Greenland are covered by thick ice sheets that cover over a million square kilometers each. The greatest thickness of the Antarctic Ice Sheet is almost 4,800 meters (over 15,000 feet, or about 3 miles).

the size of the Antarctic ice sheet, here are two facts: it contains about 90% of the world's ice and 70% of its fresh water!

Like alpine glaciers, ice sheets are capable of doing a tremendous amount of erosion. The glacial ice in ice sheets wears down and rounds obstacles that stand in its path as ice moves. They also scour out deep depressions. The Great Lakes in North America are examples of lakes that were created by the erosional work of past ice sheets.

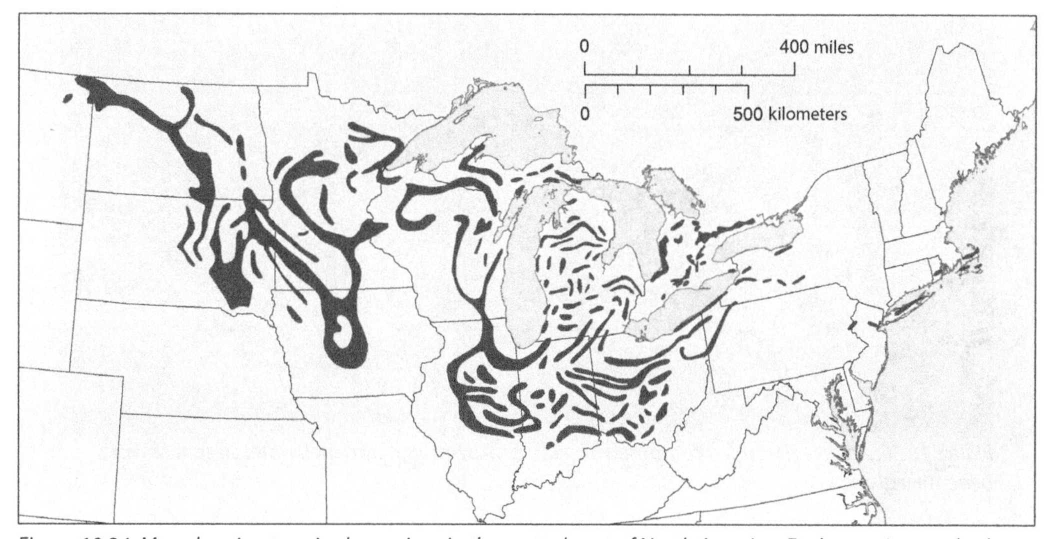

Figure 10.34. Map showing terminal moraines in the central part of North America. Each moraine marks the extent of ice sheets that moved down from northern Canada.

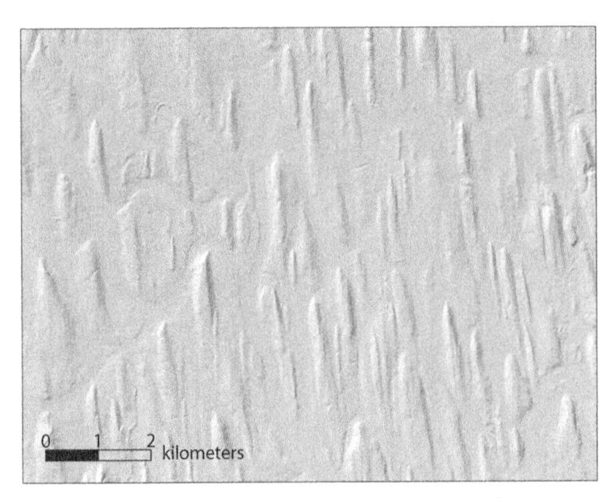

Figure 10.35. A computer-generated shaded relief map of drumlins in western New York. Drumlins are created beneath moving ice that is flowing from north to south (top to bottom).

Ice sheets make a variety of depositional landforms, including moraines. Much of the landscape of Canada and the northern United States is covered by *ground moraine*. This is till that covers the ground surface like a blanket. There are also long terminal moraines marking the southward edge of ice sheets that once covered part of North America, as shown on the map in Figure 10.34.

As ice sheets advance over previously deposited moraines, the moving ice sometimes molds material into elongated hills known as *drumlins*. The steepest end of a drumlin faces the direction from which the glacier came, as shown in Figure 10.35. Another landform that forms beneath glaciers is an *esker*—a long, winding ridge of layered sediments. Eskers, such as the one shown in Figure 10.36, form by streams that flow beneath melting

Figure 10.36. An esker is a ridge composed of sediments deposited by a meltwater stream flowing beneath a glacier.

glaciers, and typically have somewhat meandering shapes, just like normal streams. The esker ridge is composed of gravel and sand deposited in the sub-glacial stream. When the glacier is gone, all that remains is the winding ridge.

Figure 10.37. Areas once covered by ice sheets, such as shown in this topographic map of northern Minnesota, typically have numerous kettles filled with water forming lakes and swamps.

As ice sheets melt, they sometimes leave large blocks of ice on the land that become surrounded and covered by sediments from the melting glacier. When the ice blocks melt, depressions called *kettles* are formed. Kettles may be roughly circular or irregular in shape. The landscape across much of the northeastern section of the United States, from North Dakota to Maine, is covered by many thousands of kettles. Most of these are filled with water forming kettle lakes or wetlands.

Learning Check 10.3

1. How do glaciers form?
2. How do glaciers move?
3. Compare and contrast alpine glaciers and ice sheets.
4. Draw a mountain landscape that has been modified by alpine glaciers, with labels for a cirque, tarn, horn, arête, hanging valley, and U-shaped valley.
5. Describe how the following features form: terminal moraine, drumlin, esker, kettle lake.

10.4 Glaciation in Earth's past

In the next chapter of this book, we will examine Earth's past, attempting to answer questions such as "How old is the Earth?", "What is the history of Earth?", and "How do we know?" In this section, we will take a first look at Earth history by learning about the Ice Ages that occurred during the most recent 2.6 million years of Earth history.

10.4.1 Evidence for an Ice Age

For many centuries, people have observed boulders on the land surface in parts of northern Europe that seem out of place, such as the one shown in Figure 10.38. These boulders, called glacial *erratics*, are found in an area north of a line running from the British Isles to Russia, but not south of this line except in mountainous areas such as the Alps. These boulders have different compositions than local rocks, so they are not formed by weathering of rocks in place; they look like they were simply carried and dropped in place. Erratics scattered across England have the same composition as bedrock in Scotland, and those in Germany have the same composition as bedrock in Norway and Sweden. Before the early 1800s, some speculated that the erratics were carried by icebergs during Noah's flood and dropped onto the ground as the icebergs melted. This was a reasonable hypothesis at the time because mariners had observed icebergs carrying large rocks.

Figure 10.38. A glacial erratic in northern Europe, carried to its present location by glaciers over 10,000 years ago.

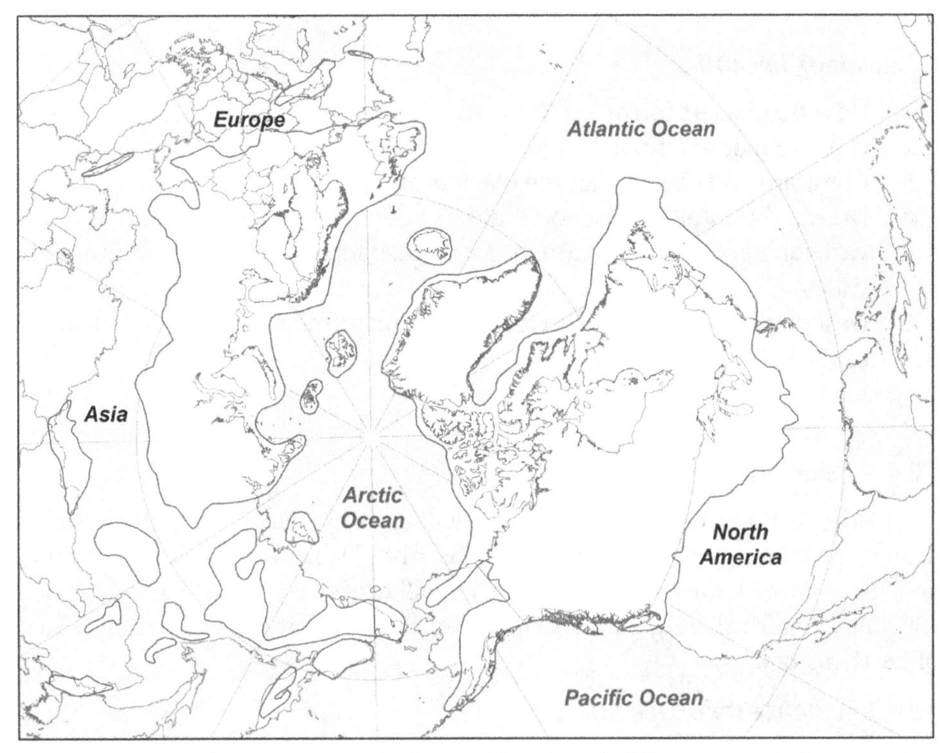

Figure 10.39. During Earth's most recent Ice Age, peaking about 20,000 years ago, glaciers covered much of North America and northern Europe and Asia. Smaller areas of glaciation existed in mountain ranges throughout the world.

As scientists learned more about how glaciers work, especially through observations of glaciers in the Alps, it became clear that a more reasonable explanation for the presence of erratics is that they were carried to their present locations by glaciers flowing over land rather than by icebergs. This was one of the first clues that vast areas of the continents were once covered by thick glaciers in a time period before human history. An *Ice Age* is a period of time during which there is a significant amount of glaciation on Earth's surface, even outside of polar areas.

In the mid-1800s, scientists discovered a number of additional clues that indicated that much of Europe and North America were once covered by vast ice sheets, and that much larger glaciers once existed in mountain ranges such as the Alps and Rockies than what are found there today. As recently as 20,000 years ago, up to one third of Earth's land surface was covered by glaciers, as illustrated in Figure 10.39.

Evidence for past glaciation includes the following:

- The presence of erratics on Earth's surface.
- Striations and grooves cut into solid rock. These not only give evidence of past glaciation, but indicate the direction ice was moving.
- Moraines. In many places, terminal moraines mark the farthest advance of glaciers. In other areas, ground moraines blanket the surface with till.
- Other depositional landforms. Areas once covered by ice sheets have features such as kettle lakes, eskers, and drumlins for which we have no other explanation than that the area was once covered by a glacier.
- Glacial landforms in mountains. The presence of many types of glacial erosional landforms, such as U-shaped valleys, cirques, hanging valleys, and arêtes, testifies to the past presence of large glaciers.

10.4.2 The Timing and Worldwide Effects of Glaciation

Our present understanding is that Earth has experienced an Ice Age for about the past 2.8 million years.[6] For much of this time, glaciers covered large areas of the continents, as indicated in Figure 10.39. But throughout this time, Earth's climate has been somewhat cyclical, with long periods of glaciation followed by shorter periods of time called *interglacials* during which the global climate was warmer and ice sheets over North America and Eurasia largely disappeared. Figure 10.40 shows how Earth's climate has fluctuated over the past 400,000 years. The last glacial period ended about 13,000 years ago and it appears that we are currently in an interglacial period. If these cycles continue, Earth will enter another glacial period in perhaps 50,000 years. These cycles of glacial and interglacial periods appear to be related to subtle changes in Earth's orbit around the sun.

Ice Age climate cycles have affected the entire Earth. One way the formation of ice sheets affects the entire planet is by changing sea level. At the peak of glaciation, the worldwide volume of ice was tens of millions of cubic kilometers. This

6 Glaciation began in Antarctica and Greenland millions of years earlier.

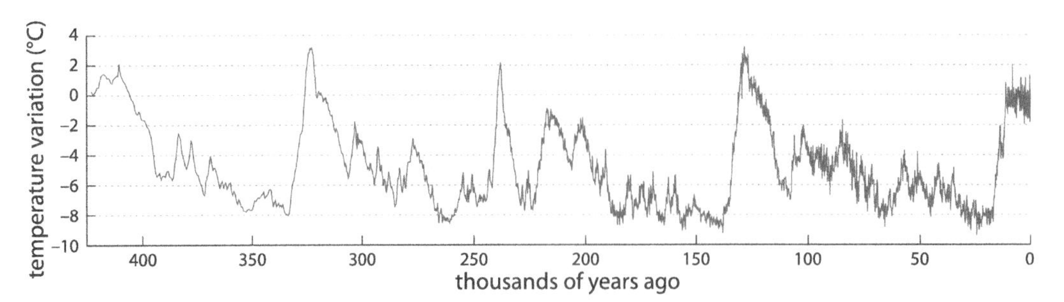

10.40. Earth's temperature variations over the past 400,000 years. The graph shows how much colder or warmer Earth's overall surface temperature has been than at present, measured in °C. High points on the line indicate interglacial times, and lower portions indicate glacial times.

tremendous volume came from seawater, which evaporated from the ocean surface and fell on the continents as snow. The result was that sea level around the globe dropped by about 120 meters (400 feet), not only exposing the Bering Land Bridge referred to at the beginning of this chapter, but large areas of the continental shelf around all the continents.

Most of the animal fossils found in Ice-Age deposits are from animals that are familiar to us, such as horses and camels—both of which lived in North America around 10,000 years ago. But there were other mammals roaming the plains of North America back then that are now extinct, such as saber-toothed cats, woolly mammoths, and mastodons. An Ice Age scene created by Spanish "paleoartist" Mauricio Antón is shown in Figure 10.41. Many of these mammals became extinct after the last glacial period. Scientists debate as to whether this extinction was due to a changing climate, overhunting by humans, or a combination of causes.

For most of its history, Earth has been warmer than at present, though there have been other Ice Ages in the past, as evidenced by features in ancient rocks such as striations and deposits that resemble till. Geologists do not completely under-

10.41. Ice Age mammals included woolly mammoths and woolly rhinoceroses.

stand why Earth has been cooling for the past several million years. One reasonable hypothesis is that as continental plates have moved through plate tectonics, they have moved into positions that are favorable for glaciation, with large portions of North America and Eurasia being in polar or subpolar latitudes, and Antarctica being centered on the South Pole. The position of the continents also affects how currents circulate through the oceans, and at present the Arctic Ocean is largely isolated from warm-water currents flowing in the other oceans.

Learning Check 10.4

1. What evidence led scientists, starting in the 1800s, to conclude that Earth has experienced ice ages in the past?
2. What makes a glacial erratic distinct from other large boulders on Earth's surface?
3. Explain how the size of ice sheets on Earth's surface affects sea level.

Chapter 10 Exercises

Answer each of the questions below as completely as you can. Write your responses in complete sentences unless instructed otherwise.

1. Select a human activity that increases the risk of mass movement, and suggest ways to reduce or eliminate the risk of that activity.
2. What is the connection between wildfires, rain, and mass movement?
3. Suggest a hypothesis as to why water in a stream can move gravel for long distances, but wind cannot.
4. Explain why sand dunes such as barchans and transverse dunes do not form in humid areas.
5. How is it possible that glaciers exist near the equator?
6. How would partial melting of the Greenland Ice Sheet affect coastal cities like Miami or New York City?
7. Investigate one of the following:
    - Mass movement risk in your state or region.
    - Wind erosion and deposition in your state or region.
    - Glacial history of your state or region. (This could include sea-level changes caused by past glaciation)

Experimental Investigation 7: Studying Glaciers with Topo Maps

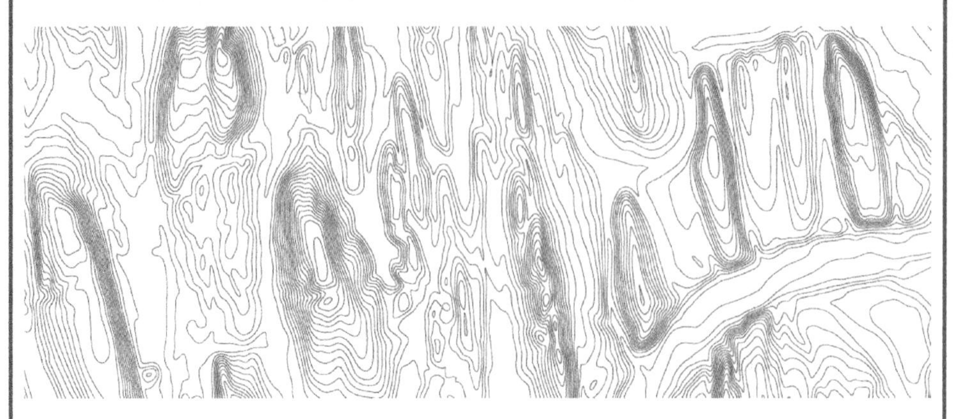

Overview

The goal of this investigation is to interpret topographic maps which show landforms created by alpine glaciers and continental ice sheets. Some landforms created by glaciers, such as drumlins, eskers, horns, arêtes, and U-shaped valleys, are large and distinctive, and can be readily identified on topographic maps. Other glacial landforms are small or subtle and don't stand out on most topographic maps. In this laboratory exercise, you will examine two topographic maps showing landforms created by alpine glaciers, and two showing landforms created by continental ice sheets.

Basic Materials List

- Four topographic maps: Logan Pass, Montana; Anchorage, Alaska; Passadumkeag, Maine; and Palmyra, New York.
- Ruler and graph paper

Map 1—Logan Pass, Montana

Glacier National Park in Montana is not named primarily because it has glaciers—though it does have a handful of small glaciers—but because of the extensive sculpting of the landscape done by glaciers in the past. Virtually every mountain, ridge, and valley in the park has been shaped by the erosional force of alpine glaciers. The hundreds of lakes in

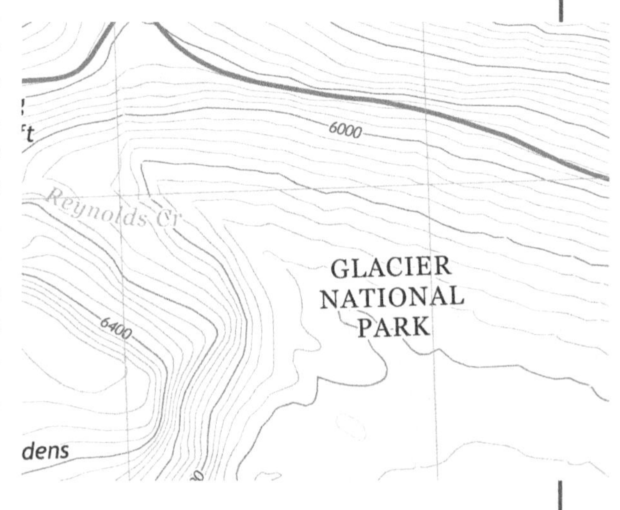

the park were almost all formed either by glacial erosion or by natural dams formed by glacial moraines.

1. What is the scale of this map? What is the contour interval?
2. This map has a number of horns, many of which are named. What are the names of three horns on this map?
3. Describe the ridges that run from Citadel Mountain to Dusty Star Mountain; and from Reynolds Mountain to Heavy Runner Mountain. Are the slopes of these ridges gradual or steep? How do you know? What type of landforms are these ridges?
4. What are the names of two tarns found on this map? What type of landforms are tarns found in?
5. There are a few small glaciers on this map. What type of landform are these glaciers found in?
6. Draw a topographic profile from Citadel Mountain to Hanging Gardens. One of the valleys is higher in elevation than the other. What landform is this higher valley? What landform separates the valleys?

Map 2—Anchorage, Alaska

The coastal mountain ranges of southern Alaska receive a high amount of precipitation, much of which falls as snow, especially at higher elevations. The result is that these mountain ranges have the largest coastal glaciers in the world outside of Greenland and Antarctica.

Contour lines on ice are shown as blue lines rather than brown lines.

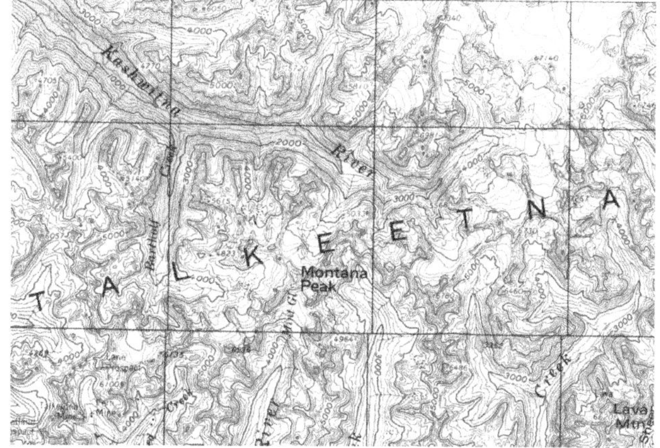

1. What is the scale of this map? Does this map cover a larger area than the Logan Pass map, or a smaller area?
2. What is the contour interval of this map? Suggest a reason why the contour interval of this map is different than the contour interval of the Logan Pass map.
3. Determine the gradient (feet per mile and percent slope) of Harvard Glacier from its end at Harvard Arm to the final "R" in "Harvard Glacier."

4. Harvard Glacier has a number of smaller glaciers flowing into it, such as Radcliffe Glacier and Lowell Glacier. If all of these glaciers were to melt, what type of landform would be seen where these smaller glaciers once were?
5. What type of landforms are Harriman Fiord, College Fiord, and Unakwik Inlet? How did these features form?
6. Describe the pattern of channels made by the Knik River, Nelchina River, and Matanuska River. What type of streams are these rivers?
7. What evidence do you see on this map that glaciers were once more widespread than they are now?
8. Propose a hypothesis (or multiple hypotheses) for why glaciers are much larger on the Chugach Mountains in the southern part of the map than they are on the Talkeetna Mountains in the northern part of the map.

Map 3—Passadumkeag, Maine

The state of Maine, along with the rest of the New England States, was completely covered by an ice sheet prior to about 12,000 years ago. Mountains and hills in the northern part of the state have mostly been rounded off by the relentless movement of debris-carrying ice. Lower areas, such as on most of the area on this map, are blanketed by thick deposits of glacial till. Landforms in these lower areas include kettles, many of which contain ponds; moraines, swamps, and eskers.

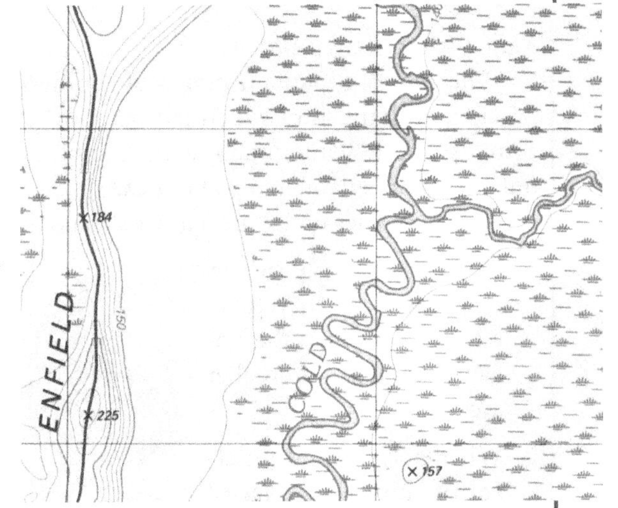

1. What is the scale of this map? What is its contour interval?
2. Find the Enfield Horseback, a long ridge that runs north-south the whole length of this map. What type of glacial landform do you think this is? Why? How does this type of landform form?
3. What type of material is the Enfield Horseback made of? There is a clue on the map near where Caribou Road crosses the ridge.
4. What other signs of glaciation do you see on this map?

Map 4—Palmyra, New York

Most of New York was also covered by an ice sheet during the Ice Ages. This map has very distinctive glacial landforms.

1. What is the scale of this map? What is its contour interval?
2. This map has a large number of elongated hills. On average, how long and wide are the hills, in feet?
3. What is the average height of the hills, in feet?
4. What landform are these hills? Are they erosional or depositional landforms?
5. What direction was the ice flowing? How do you know?

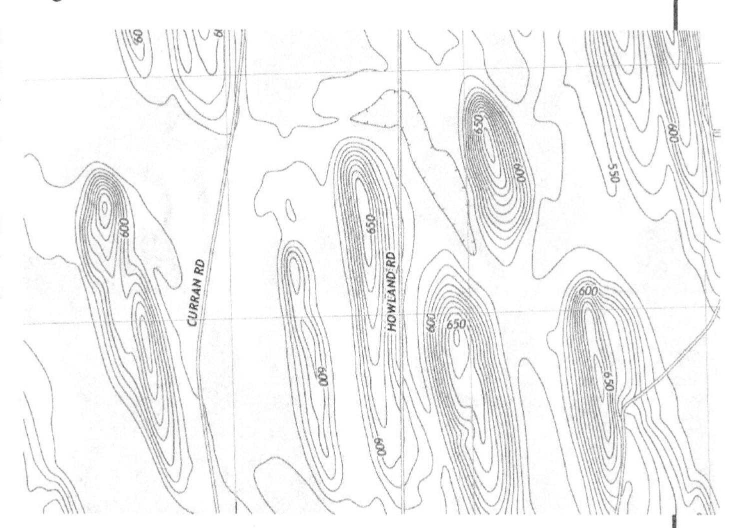

Chapter 11

Unraveling Earth History

If you go to a natural history museum, you will likely see displays depicting what scientists think Earth looked like at various times in its past. The diorama shown above would be labeled something like "Cambrian Seafloor, 510-540 million years ago." Some displays show ancient seafloors, and others show what Earth's land surface might have looked like tens or hundreds of millions of years ago. If you examine these dioramas closely, you notice that the ones from older periods of Earth history contain many strange and unfamiliar creatures, and that the scenes become increasingly familiar as the past approaches the present.

How do Earth scientists know what Earth was like at various times in the past? How do they determine numbers like "540 million years ago?" In this chapter, we explore how geologists reconstruct the order of events in Earth history, how they determine how old certain rocks are, and why they think certain rocks were formed in different types of environments, such as in a warm, shallow sea as pictured above. We conclude the chapter with a short overview of how Earth has changed over time, from its beginning as a chaotic, lifeless world over four billion years ago to a planet that is ideally suited for the flourishing of its inhabitants—including humans—today.

Objectives

After studying this chapter and completing the exercises, you should be able to do each of the following tasks, using supporting terms and principles as necessary.

1. Distinguish between physical geology and historical geology.
2. Describe why scientists such as Hutton and Lyell concluded that Earth must be millions of years old.
3. State the modern understanding of the principle of uniformitarianism.
4. Explain the difference between relative-age dating and absolute-age dating.
5. Use the principles of original horizontality, superposition, and cross-cutting relationships to determine the relative-age of rock units.
6. Define *unconformity* and describe how three different types of unconformities form.
7. Describe three ways fossils form.
8. Explain how the isotopes of an element differ from each other.
9. Describe how the concepts of radioactive decay and half-life are used to determine the age of rocks.
10. Describe how different sedimentary environments produce sediments with different characteristics.
11. Describe the principle of fossil succession.
12. Give a brief overview of how Earth has changed over geologic time.
13. Give a brief overview of how life on Earth has changed over geologic time.

Vocabulary Terms

You should be able to define or describe each of these terms in a complete sentence or paragraph.

1. absolute-age dating
2. amber
3. Cambrian explosion
4. cast
5. cross-bedding
6. depositional environment
7. eon
8. epoch
9. era
10. formation
11. fossil
12. geologic column
13. half-life
14. historical geology
15. isotope
16. mass extinction
17. mold
18. nuclide
19. original preservation
20. period
21. petrification
22. physical geology
23. principle of cross-cutting relationships
24. principle of fossil succession
25. principle of original horizontality
26. principle of superposition
27. radioactive decay
28. radiometric dating
29. relative-age dating
30. trace fossil
31. unconformity
32. uniformitarianism

11.1 Geologic Time

Much of what you have learned in this course up to this point has involved Earth processes that are occurring at present, such as volcanism, earthquakes, weathering, erosion, and the actions of water, wind, and ice on Earth's surface. These are all topics of *physical geology*, which is the study of the materials Earth is made of and the processes that act on those materials. Now we can turn to the topic of *historical geology*—the study of Earth's past. You have already learned some things about

Earth history since many of the processes we have covered so far involve time. In Chapter 6, you read about plate tectonics and learned that Earth's continents were once assembled into a supercontinent known as Pangea, that at some distant time in Earth's past Pangea broke into smaller pieces, and that these pieces move across Earth's surface at rates of a few centimeters per year. In Chapter 10 you read about the ice age that has gripped Earth for much of the past 2.6 million years. In this chapter, you will learn more, not only about Earth's history, but also about how geologists reconstruct that history.

11.1.1 The Development of the Concept of Geologic Time

In ancient times, philosophers such as Aristotle, Plato, and many others believed that the universe was eternal—that our planet had always existed in the past and would always exist in the future. This contrasted with the explanation given in the biblical book of Genesis, which opens with, "In the beginning, God created the heavens and the earth." Now scientists understand that there is clear geological evidence telling us that Earth has not existed forever, but came into existence at some time in the past. If Earth had a beginning, how long ago did that beginning occur, and how did it happen?

In the middle ages in Europe, most scholars accepted a very literal reading of Genesis, which to them implied that Earth was about 6,000 years old. This age was calculated by adding the ages of various people recorded in the genealogies in Genesis, such as Noah, Abraham, and Isaac. One scholar, James Ussher, an archbishop in Ireland, determined that Earth must have been created in 4004 BC. A number of Biblical scholars still hold to this young-Earth interpretation. Many other Bible-believing scholars, however, from the 1800s until the present, have rejected the young-Earth interpretation of Genesis. Some of these biblical experts believe each day in Genesis 1 represents long periods of time. Others believe that it is not possible to tie the description in Genesis 1 to events in Earth history.

The sciences of astronomy and physics developed significantly in the 1500s through 1700s through the efforts of scientists such as Copernicus, Brahe, Kepler, Galileo, and Newton. Detailed study of Earth's rocks and surface, however, did not begin until the 1700s. By the end of the 1700s, many scientists had come to the conclusion that Earth must be considerably older than 6,000 years. Some suggested that Earth might be tens of thousands of years old, while others proposed that Earth is really millions of years old.

11.1.2 Hutton and Lyell

James Hutton, a Scottish farmer, lived in the late 1700s. He is considered to be one of the founders of the modern science of geology. Hutton, portrayed in Figure 11.1, recognized that rocks exposed to Earth's atmosphere weather to produce gravel and soil. He also observed that many layers of solid rock contained what appeared to be the same sorts of material—sand and gravel derived from preexisting rocks. Hutton also observed that sediments were still being deposited at the present time, primarily in the sea. These observations led Hutton to a view of

Earth history in which destruction of older rocks was balanced by the creation of new rocks. This implies that the record of the past recorded in rocks could be explained by processes operating on Earth today. By observing that the rates of weathering, erosion, and deposition typically occur at slow rates, and by considering the amount of sedimentary rocks that exist in Earth's crust, Hutton and other scientists of his time concluded that Earth must be much older than just a few thousand years. It became clear to these scientists that too many events had happened in Earth history to have occurred in such a short amount of time.

Figure 11.1. James Hutton in 1776. Hutton was an important early geologist who recognized that Earth appeared to be more ancient than was widely accepted at the time.

Hutton's ideas were further developed by Charles Lyell, a British geologist in the early to mid-1800s. Lyell's principle of *uniformitarianism* states that the processes occurring today—mountain building, erosion, deposition of sediments, earthquakes, and so on—also occurred in the past. Rocks created by volcanic eruptions in the past are like rocks created by present-day volcanoes. Streams on the ancient Earth left deposits that look like the deposits of modern streams. Waves on a beach followed the same laws of physics a million years ago that they do today. Figure 11.2 shows a painting of Lyell, and Figure 11.3 is a remarkably accurate geologic diagram from his textbook on geology.

Figure 11.2. Charles Lyell in 1840. One of Lyell's many contributions to the science of geology was the principle of uniformitarianism. This principle holds that Earth's geologic features were produced by processes that follow the same laws of nature that we can observe in the world today.

The idea that the laws of nature are uniform—that is, that they don't vary from one time and place to another—is consistent with the Christian view that God is not only the creator of the material that the universe is made of, but the lawmaker of the universe as well. God didn't just create the planets and stars; he created the laws that govern their motions—laws that are understandable to humans. The same is true of the processes studied by Earth scientists such as the crystallization of minerals from magma; weathering, erosion, and deposition of sediments; and the formation and melting of glaciers. Understanding how nature works—the laws

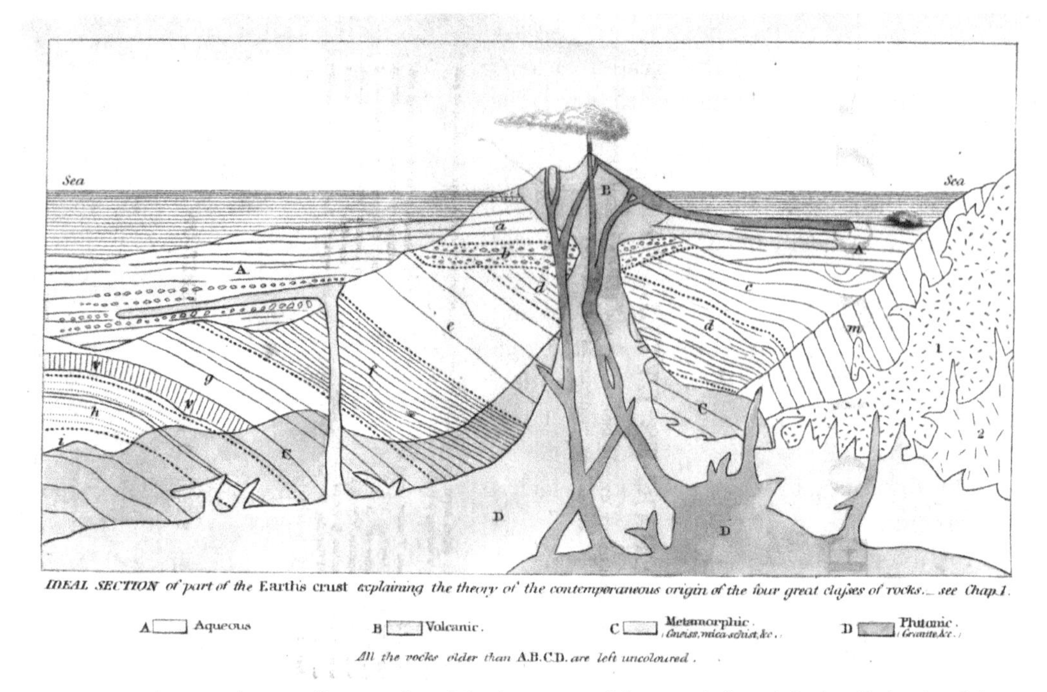

Figure 11.3. A diagram from Lyell's Principles of Geology, *one of the most influential scientific books of the 1800s. The diagram shows both intrusive and volcanic igneous rocks, tilted and horizontal sedimentary rocks, and zones of metamorphism.*

put in place by God in the beginning—also helps us to decipher Earth's history from rock layers.

The principle of uniformitarianism does not mean that everything that has ever happened must be explained by natural laws and processes. As Christians we believe that there have been events in history that have occurred outside of natural laws. We are told in the Bible that God created the universe from nothing,[1] a creation event that included the creation of the laws that govern the universe. Miracles—such as the parting of the Red Sea in Exodus, Jesus healing people and feeding the 5,000, and the Father raising Jesus from the dead—cannot be explained by the actions of natural laws, but this does not mean that these events did not happen.

Geologists today do not have as rigid a view of uniformitarianism as Lyell did. Geologists recognize that though the laws of nature have not changed over the millions of years of Earth history, the intensity of Earth processes can change dramatically. While Lyell believed that all geological processes were slow and steady, geologists today recognize that catastrophic events also have been important in Earth history. These catastrophes range from local events such as landslides and floods to events that are of global importance, such as the impact of large meteorites on Earth's surface and the eruption of huge volcanoes on a scale never witnessed by humans.

1 Genesis 1:1, John 1:1-3, Hebrews 11:3

By the middle of the 1800s, most geologists, many of whom were Christians, accepted the idea that Earth is millions of years old. Many Bible scholars and theologians also accepted this view of an ancient Earth. It might be helpful to you at this point to review Section 3.4.3, which gives a brief introduction to thinking about the relationship between Earth science and Christianity.

Learning Check 11.1

1. Distinguish between physical and historical geology, and give two examples of topics that would be investigated in each.
2. Explain what is meant by the term *uniformitarianism*.

11.2 Relative Age

In the 1800s, geologists were able to reconstruct the order in which geologic events happened during Earth history, even though they did not have the ability to determine precise ages for those events. *Relative-age dating* is the process of determining the order of events in Earth history.

11.2.1 Rock Layers

Much of what we know about Earth history has been determined through the study of widespread layers of rocks. Some of these layers are very large—tens or even hundreds of meters thick and extending across areas measured in millions of square kilometers. Other layers have thicknesses measured in millimeters. Most layers are made of sedimentary rocks, but volcanic rocks can make layers too, such as the extensive layers of lava flows present on the Columbia Plateau of the northwestern United States, shown in Figure 11.4.

Geologists lump similar rock layers into units called *formations*. Formations are sets of layers that have distinct properties, such as the type of rock or types of fossils found in the layers. Figure 11.5 shows four formations that make up the top layers of the rocks exposed in the Grand Canyon.

Figure 11.4 Layers composed of basalt, a volcanic rock, in eastern Washington.

Figure 11.5. The four formations at the top of the Grand Canyon. These formations are continuous for long distances, and always occur in the same order.

- The Kaibab Formation primarily consists of layers of fossiliferous limestone and is resistant to erosion, so it forms cliffs.
- The Toroweap Formation makes a more gradual slope beneath the Kaibab, and is composed of more easily-eroded layers of shale and gypsum, with a few layers of sandstone that are more resistant to erosion.
- The Coconino Formation is composed of thick layers of cliff-forming white sandstone.
- Beneath the Coconino is the Hermit Formation, which has eroded into a slope and is composed of reddish-brown shale.

These formations are found throughout large areas of the southwestern United States and are always found in the same order. The sedimentary rock formations found in other places are different because each part of Earth has a unique geologic history. To geologists, rock layers are like the pages of a book, with each page telling a new part of the story of Earth's past. We will now look at some principles that geologists use to read those pages.

11.2.2 Principles of Relative-Age Dating

Geologists use a set of principles, or rules, to help them interpret the story of Earth's past. We have already seen one of these principles, the principle of uniformitarianism. This principle tells us that the natural laws we observe in the world at the present time also operated in the past. Another important rule is the *principle of original horizontality*. This principle states that sedimentary layers are deposited in horizontal or near-horizontal layers. The layers at the top of the Grand Canyon,

Figure 11.6. These layers in a roadcut near Denver, Colorado were deposited horizontally and were tilted as the Rocky Mountains were uplifted.

shown in Figures 8.16 and 11.5, are still in their original horizontal positions. If layers are observed to be tilted at present, such as the layers in Figure 11.6, we know that the layers have been deformed and moved from their original horizontal positions.

The next important rule for determining relative age is the *principle of superposition*. The principle of superposition states that in a stack of sedimentary rock layers that have not been severely deformed, the oldest layers are at the bottom and the youngest layers are at the top. The oldest of the four formations in Figure 11.5 is the Hermit Formation. The next oldest is the Coconino Formation, which was deposited on top of the Hermit. The Toroweap was deposited next, and the Kaibab Formation is the youngest. Exceptions to the principle of superposition are caused by mountain building, when older rocks are moved above younger rocks through faulting or folding. Figure 6.33 illustrates how older rocks were thrust over younger rocks in Glacier National Park.

Another important rule for determining relative age is the *principle of cross-cutting relationships*: features that cut through existing rock layers are always younger than the rocks they cut across. In Figure 11.7, we know that rock layers A through G are older than the reverse fault that cuts through them. In other words, the fault could not possibly be older than the rocks that it cuts; the rock layers were there before the fault developed. The

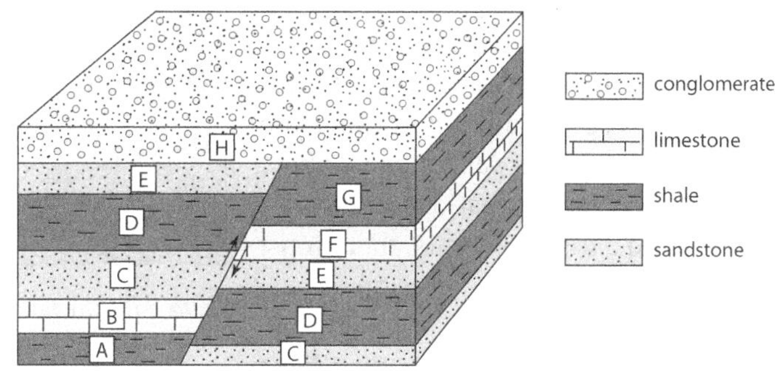

Figure 11.7. The reverse fault is younger than the rocks it cuts, but older than the top layer, which it does not cut.

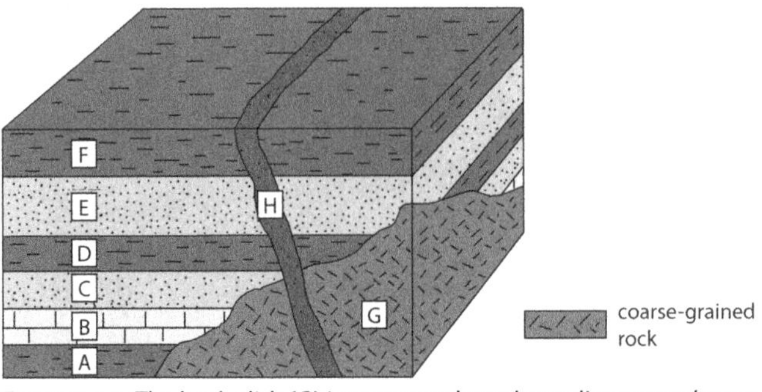

Figure 11.8. The batholith (G) is younger than the sedimentary layers (A-F) it intruded. The dike (H) is younger than all rocks it cuts.

top layer, H, has not been cut by the fault. Thus we know that layer H must have been deposited after the fault was formed, and that the fault is no longer active. The principle of cross-cutting relationships applies to igneous intrusions such as batholiths and dikes as well. In Figure 11.8, the igneous rocks must be younger than the sediments they intrude.

11.2.3 Unconformities

An *unconformity* is an erosional surface preserved in a stack of rock layers. Unconformities represent periods of time when erosion was occurring on Earth's surface rather than deposition of sediments. There are three types of unconformities, illustrated in Figure 11.9. The first is the *nonconformity*, where layered rocks lie on top of non-layered rocks. The existence of a nonconformity suggests that the underlying igneous or metamorphic rocks were uplifted and eroded before deposition of horizontal sedimentary rocks.

A second kind of unconformity is the *disconformity*, where horizontal sedimentary rocks lie on top of other horizontal sedimentary rocks. Disconformities can be rather subtle, especially when seen from a distance. In order for the contact between two formations to be considered a disconformity, there must be some evidence of erosion or passage of a considerable time period. Disconformities often display signs of weathering and erosion, such as

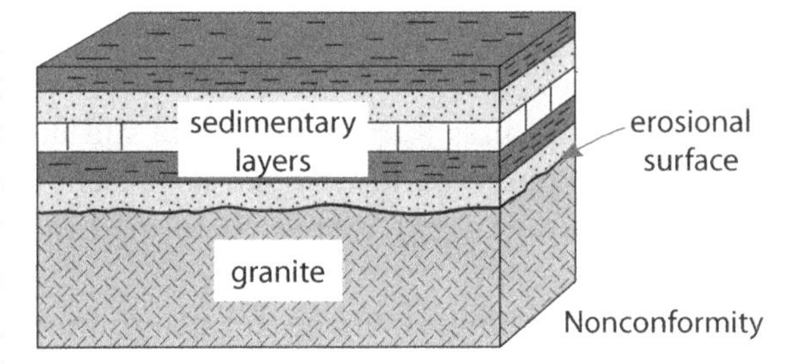

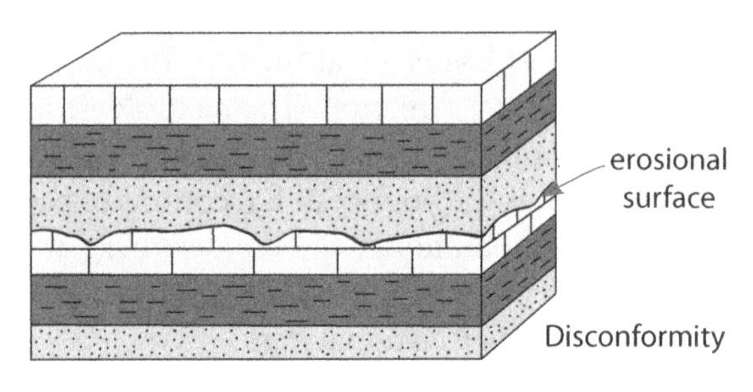

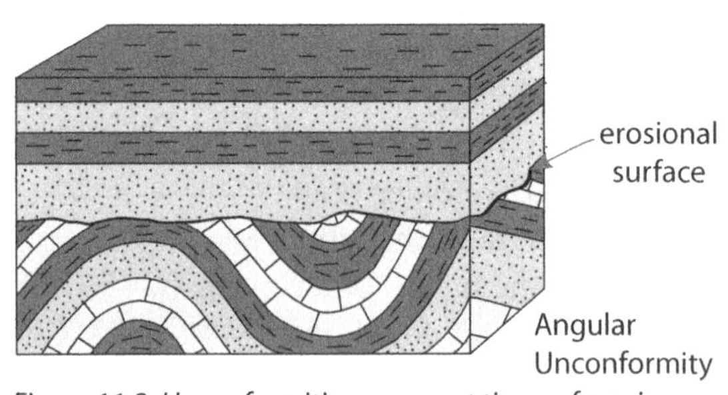

Figure 11.9. Unconformities represent times of erosion rather than deposition during Earth history. Three types of unconformities are nonconformities (top), disconformities (center), and angular unconformities (bottom).

the remains of ancient soils, animal burrows, and stream channels cut into the underlying rocks. Figure 11.10 shows one of several disconformities that exist within the horizontal layers of the Grand Canyon.

Figure 11.10. The disconformity at the top of the Muav Limestone in the Grand Canyon consists of ancient channels eroded into the underlying limestone. The channels were filled with the sediments of the Temple Butte Formation before the deposition of the Redwall Limestone.

Figure 11.11 illustrates a third type of unconformity, the *angular unconformity*. An angular unconformity is created when sedimentary rocks are uplifted and tilted, and then weathered and eroded to create a horizontal surface. New layers of sediments are then deposited on top of the erosional plane.

Figure 11.11. An angular unconformity in Scotland that was studied by Hutton in the 1700s. The layers closer to the camera are tilted nearly vertically, while the upper layers are gently tilting.

11.2.4 Putting it all Together

Many areas have complex geologic histories involving periods of volcanism, igneous intrusion, mountain building, deposition of sediments, metamorphism, and erosion. Geologists can apply the principles outlined above—original horizontality, superposition, cross-cutting relationships, and unconformities—to decipher the story recorded in rocks.

Figure 11.12 shows a more complex geological diagram including intrusive igneous rocks (A), volcanic igneous rocks (B), folded sedimentary rocks (C–I), and horizontal sedimentary rocks (J–L). Using the principle of superposition, the deepest sedimentary rocks are the oldest sedimentary rocks, and the shallowest are the youngest. There is an angular unconformity beneath layer J, representing a prolonged period of erosion. The large igneous intrusion, A, cuts across the folded sedimentary layers, so layers C–I must be older than A. The smaller igneous body, B, cuts across intrusion A, folded layers C–I, and horizontal layer J. Using the principles outlined in this chapter, the geologic history of this area looks like this:

- Deposition of sedimentary layers C through I.
- Mountain building and folding of layers C through I.
- Intrusion of A, which is part of a batholith.
- Erosion of folded layers (C–I) and batholith (A).
- Deposition of layer J. The contact between the folded layers C–I and horizontal layer J is an angular unconformity. The contact between the batholith (A) and horizontal layer J is a nonconformity.
- Formation of B, part of which is a dike, and part of which is a lava flow that flowed over layer J.[2]
- Deposition of sedimentary layers K and L.

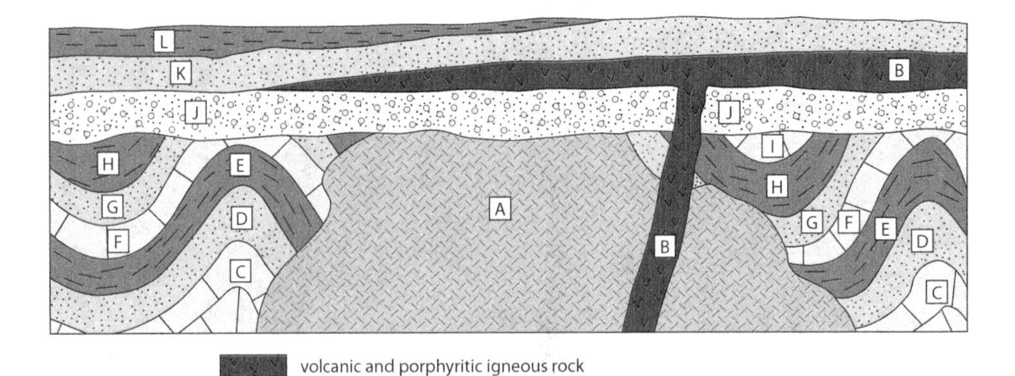

Figure 11.12. A diagram of a portion of Earth's crust with a more complex history.

2 Alternatively, B could be interpreted as a dike and sill that intruded between J and K after J and K were deposited.

Learning Check 11.2

1. Explain why the rock layers in Figure 11.12 do not violate the principle of original horizontality.
2. Explain why an igneous intrusion, such as a dike or batholith, cannot be older than the rocks it cuts.
3. Describe how a disconformity forms.
4. Describe how an angular unconformity forms.

11.3 Fossils

Fossils are the remains or traces of once-living organisms that are preserved in the geologic record. When you think of fossils, you may think of impressive dinosaur specimens in a museum, such as the one in Figure 11.13, but most fossils are those of invertebrates[3] such as clams, corals, snails, and starfish that lived in ancient seas.

11.3.1 Types of Preservation

Most organisms quickly rot or are eaten after they die. When an animal dies, its flesh may be eaten by scavengers in a few short days; what is left is broken down by bacteria and fungus. After a while, the only thing left will be the hard parts, such as

Figure 11.13. Dinosaurs in a group called the sauropods were the largest animals ever to live on land. This Apatosaurus is on display in the Carnegie Museum in Pittsburgh.

3 An invertebrate is an animal without a backbone.

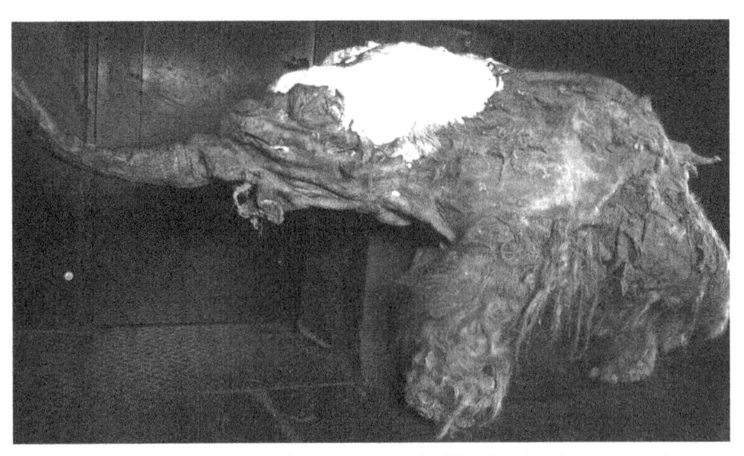

Figure 11.14. This young woolly mammoth died in Siberia about 39,000 years ago and was preserved in permafrost (permanently frozen soil).

shells or bones, and even these will decompose over time. This decomposition is part of God's design for the world; if the remains of dead plants and animals were not recycled in this fashion, Earth would quickly become covered with dead organisms. It takes special circumstances, therefore, for an organism to become preserved as a fossil. In most cases (but not all), an organism must be buried quickly by sediments in order to be preserved as a fossil. In most situations, only hard parts of organisms, such as bones, teeth, and shells, are preserved in the fossil record.

In rare cases, complete organisms, including soft parts, are preserved as fossils, a condition known as *original preservation.* In order to preserve soft parts of organisms, there must be some way of preventing the soft parts from being eaten or decomposed. One way this has happened is by freezing. In polar areas such as northern Alaska, Canada, and Siberia, scientists have unearthed Ice Age animals such as the woolly mammoth shown in Figure 11.14. Such fossil animals can be complete with thick fur, inner organs, and even their last meals preserved in their stomachs. The bodies of the mammoths must have been frozen shortly after they died, must have been buried quickly by sediments, and then remained frozen for thousands of years before being uncovered. Other ways in which organisms can be preserved intact include drying out (*dessication*) in desert environments, and preservation in

Figure 11.15. A tiny gnat almost perfectly preserved in amber (fossilized tree resin).

hardened tree resin called *amber*. Figure 11.15 shows an insect preserved in amber.

Figure 11.16. Logs of petrified wood at Petrified Forest National Park, Arizona.

Much more commonly, fossils are preserved by some sort of alteration of hard parts, a process sometimes referred to as *petrification*. An example of this is petrified wood, such as shown in Figure 11.16. Petrified wood is formed when wood becomes buried, and mineral-rich groundwater passes through the wood. Pore spaces in the wood become filled with minerals such as silica, and eventually the woody material is completely replaced by minerals. Ultimately, the wood is transformed into solid rock. The same thing happens to bones. Most bones preserved in the fossil record are not made of the original bony material that was in the animal, but of mineral material that has replaced bone in a process similar to the formation of petrified wood. Most fossils of marine organisms, such as oysters, snails, and corals, have hard parts that are easily preserved. At a microscopic level, the calcium carbonate hard parts of these organisms have been altered, but the fossil itself retains the shape of the original organism.

A widespread means of fossil preservation is the creation of *molds* and *casts* of organisms. If you were to take a clam shell and press it into the sand on a beach, the impression would be a mold. If you remove the shell and fill the impression with additional sediment, the infilling of the mold is a cast. A cast

Figure 11.17. Mold of a bivalve, a marine mollusk such as a clam. A mold is an impression in sediment; the shell itself is not preserved.

forms a replica of the original shell. Figure 11.17 shows a mold of a bivalve, a group of organisms that includes clams and scallops.

11.3.2 Trace Fossils

A *trace fossil* is an indirect piece of evidence of past life, such as footprints, animal burrows, and feeding patterns on the seafloor. Many types of organisms leave tracks, such as the dinosaur footprints shown in Figure 11.18. By studying trace fossils, paleontologists have been able to determine how certain dinosaurs walked, how heavy some species were, that certain types of dinosaurs traveled in groups, and that—for some species—young dinosaurs traveled with adults. Fossil footprints were created not only by dinosaurs, but by amphibians, mammals, birds, and even insects and spiders. Burrows, such as the underground passageways made by mammals, other vertebrates, or invertebrates, are also preserved in the fossil record as trace fossils. Sediments on the seafloor are home to a variety of burrowing creatures, such as marine worms. Body fossils of worms themselves are extremely scarce in the fossil record because worms have no easily preserved hard parts, but the signs of their digging through seafloor sediments are abundant in some sedimentary rocks.

Figure 11.18. Dinosaur footprints are an example of trace fossils. These tracks of various species are preserved at Dinosaur Ridge near Denver, Colorado.

Organisms leave other types of trace fossils in the rock record as signs of their past existence. One of the more interesting of these are coprolites, illustrated in Figure 11.19. Coprolites are fossilized dung or waste material. Scientists are able to examine the inside of these coprolites to determine what prehistoric animals had for dinner.

Figure 11.19. These coprolites—fossilized dung—contain bone fragments, such as the deer toe bone seen near the center, indicating the dung must have come from a carnivorous (meat-eating) animal.

11.3.3 Uses of Fossils

Fossils give scientists clues about both Earth's geologic history and the history of life on Earth. The existence of fossils of tropical plants in Antarctica tells us that the climate of Antarctica was once very different than it is now. Likewise, we know that the climate of northern North America was once warmer because paleontologists have found fossils of alligators in Canada. Many sedimentary rocks in the Earth's major mountain ranges contain marine fossils, indicating that the sediments that make up these rocks formed in ancient oceans and were later uplifted.

Knowing the type of environment in which sedimentary rocks formed is important to geologists who search for oil and gas deposits, and fossils are an important part of determining those environments. A coral-bearing limestone, for example, is likely to have high porosity, which makes it a potential reservoir rock for petroleum.

Learning Check 11.3

1. Why does it take unusual conditions to preserve the soft parts of animals in the fossil record, while hard parts such as teeth or shells are more easily preserved?
2. Explain how bones or wood might become petrified.
3. Describe three examples of types of trace fossils.
4. In what ways can fossils tell us about what Earth was like in the past?

11.4 Absolute Dating

Relative-age dating only allows geologists to determine the order in which events occurred. Various techniques of *absolute-age dating* allow scientists to determine the actual age of rocks and other geological materials. These times are measured in terms of thousands of years (ka), millions of years (Ma), or even billions of years (Ga). Scientists have a number of ways of determining the absolute age of geologic materials. In this section, we will look at radiometric dating, which is the most common method of absolute-age dating.

11.4.1 Isotopes and Radioactivity

In Chapter 4 and in previous science courses, you have learned about elements and atoms—the building blocks for all substances on and in the Earth. You learned that each atom has a nucleus made of protons with a positive electrical charge and neutrons with no electrical charge. Each element, such as hydrogen, oxygen, or sodium, has a fixed number of protons in the nuclei of its atoms, as shown back in Table 4.1. The number of neutrons in the nucleus, however, can vary to some degree. For example, carbon is element number six on the periodic table of the elements (see Figure 4.3), so atoms of carbon have six protons in their nuclei. Carbon atoms in nature, however, can have either 6, 7, or 8 neutrons. *Isotopes* are atoms

Nuclide	Protons	Neutrons	Stability	Half-Life (years)	Daughter Nuclide
hydrogen-1	1	0	stable	—	—
hydrogen-2	1	1	stable	—	—
hydrogen-3	1	2	radioactive	12.3	helium-3
carbon-12	6	6	stable	—	—
carbon-13	6	7	stable	—	—
carbon-14	6	8	radioactive	5730	nitrogen-14
potassium-39	19	20	stable	—	—
potassium-40	19	21	radioactive	1.3 billion	argon-40 calcium-40
potassium-41	19	22	stable	—	—
uranium-235	92	143	radioactive	704 million	lead-207
uranium-238	92	146	radioactive	4.5 billion	lead-206

Table 11.1. Table of isotopes for a few elements.

of the same element that have different numbers of neutrons. Isotopes are named according to their total number of protons plus neutrons, as shown in Table 11.1.

The general term for any isotope of any element is *nuclide*. Most isotopes are stable and don't ever change into different nuclides. But the nuclei of some isotopes are unstable and undergo *radioactive decay*. Decay is a process by which the atoms of elements spontaneously change into atoms of different elements. (When atoms decay, they emit high-energy particles from their nuclei. This process is called *nuclear radiation*.) As indicated in Table 11.1, the carbon-12 and carbon-13 isotopes are stable, but carbon-14 is radioactive, so atoms of carbon-14 transform into a different nuclide over time. The nuclide carbon-14 decays to is nitrogen-14. The original, radioactive isotope (in this case, the carbon-14 isotope) is referred to as the *parent nuclide*, and the new nuclide (nitrogen-14) is referred to as the *daughter nuclide*.[4]

11.4.2 Radiometric Dating

Radioactive decay occurs at rates that have been accurately measured in laboratory studies. The decay rates are used in some cases to determine the age of certain kinds of rocks and other geologic materials. The use of nuclides and radioactive decay to determine the age of materials is called *radiometric dating*. The rate of decay for an isotope is often expressed as a *half-life*. The half-life is the amount of time that it takes for half of the atoms in a radioactive sample to decay into its

4 Technically, the daughter nuclide is the result of a single decay step. The decay of carbon-14 to nitrogen-14 occurs in a single step, so nitrogen-14 is the daughter of carbon-14. However, the decay of uranium-235 to lead-207 requires a series of many decay steps with other nuclides formed in between. Lead-206 is the end of the process and is sometimes called a *granddaughter nuclide*.

daughter isotope. The longer the half-life, the slower the rate of radioactive decay.

Table 11.2 illustrates how half-life works. If a sample starts with 64 radioactive atoms (red), these atoms will decay over time until all of them are converted into atoms of the daughter isotope (yellow). After one half-life, half of the parent isotope will be remaining. After two half-lives, only one quarter of the original material will be undecayed, and so forth. This table starts with only 64 atoms; geological samples contain many millions of atoms. If a sample starts with a million atoms of a radioactive isotope, then after one half-life there will be about 500,000 of those atoms remaining. These may sound like large numbers, but accurately and precisely counting atoms is a difficult process, and sophisticated equipment is needed, such as the mass spectrometer shown in Figure 11.20.

Figure 11.20. This sophisticated mass spectrometer is used for carbon-14 radiometric dating.

No. of Half-Lives Passed	Amount of Parent Left	Illustration
0	all	
1	1/2	
2	1/4	
3	1/8	
4	1/16	
5	1/32	
6	1/64	

Table 11.2. Half-lives. After each half-life, the amount of parent material is cut in half, and the amount of daughter material increases. The red circles represent parent atoms, and the yellow circles represent daughter atoms.

Radiometric dating works something like a stopwatch that operates in reverse. Scientists can measure the current ratio of parent and daughter isotopes, and they know the rate at which the parent has decayed to produce the daughter, so they can reason backwards to determine the age of the sample. If a sample has an equal amount of parent and daughter isotopes, it should be about one half-life old. If the rock has 1/16 parent and 15/16 daughter isotopes, then it is about four half-lives old.

There are many complications that scientists have to deal with as they perform radiometric dating. For example, the original sample may already have contained some of the daughter element, or groundwater may have added or removed either parent or daughter isotopes. In some cases, scientists have figured out ingenious ways to determine how much daughter element was already in the rock when the rock formed. In other cases, these complications make it difficult or impossible to determine accurate ages for rocks.

11.4.3 Types of Radiometric Dating

Geologists have devised many different types of radiometric dating, each based on a specific parent-daughter nuclide pair. As Table 11.1 indicates, each radioactive nuclide has a distinctive half-life. Nuclides with relatively short half-lives (such as carbon-14) are useful for dating younger materials and nuclides with long half-lives are used for dating very old materials.

Most radiometric dating methods are used to determine absolute ages for igneous or metamorphic rocks. For igneous rocks, the age determined by radiometric dating will be the age at which the rock crystallized from a magma or lava. For metamorphic rocks, the age determined will be the time of metamorphism. Most sedimentary rocks cannot be dated by radiometric techniques.

Uranium-238 is a radioactive isotope that decays in a series of steps to produce lead-206, which is stable. The half-life of uranium-238 is about 4.5 billion years, so the uranium-lead method of radiometric dating is used to measure ages of very old samples. Rocks less than about 10 million years old do not contain enough daughter lead to be accurately measured, so these "young" rocks cannot be dated by uranium-lead dating.

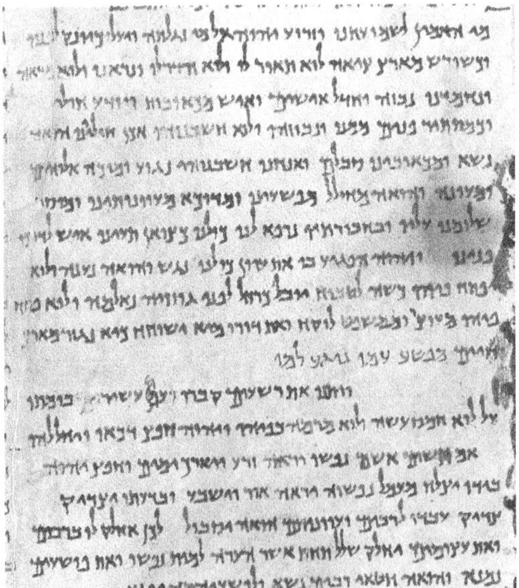

Figure 11.21. This is part of the oldest scroll known containing the Old Testament book of Isaiah. It was written on animal skin, which has been dated by carbon-14 dating. Archeologists believe this scroll was created around 125 BC.

Another common nuclide used in radiometric dating is potassium-40, which decays to form argon-40 and calcium-40 with a half-life of about 1.3 billion years. Geologists rarely measure calcium-40, but variations of potassium-argon dating are commonly done. Argon is a gas which makes up about 1% of Earth's atmosphere. This atmospheric argon probably came mostly from the decay of potassium-40 in Earth's crust. In unweathered minerals in Earth's crust, however, the daughter argon-40 produced through radioactive decay remains locked inside crystal structures. When rock samples are heated in a laboratory, the argon is released, and is measured to determine the absolute age of the rock.

Carbon-14 dating works somewhat differently than most other types of radiometric dating. Rather than dating rocks, carbon-14 dating is used to determine the absolute ages of plant and animal samples, such as archeological samples, charcoal, bones, shells, or plant debris plants buried by volcanic deposits. If a sample does not contain once-living material, it cannot be dated using carbon-14 dating.

A tiny proportion of carbon in the carbon dioxide in Earth's atmosphere is radioactive carbon-14. Plants take in this carbon-14 during photosynthesis, and animals, including humans, take in radioactive carbon-14 by eating plants. Once a plant or animal dies, its remains no longer take in carbon-14. The longer an organism has been dead, the less carbon-14 remains. Carbon-14 dating is based on the fact that the older a carbon-bearing sample is, the less carbon-14 it contains. (The amount of carbon-14 in the atmosphere is nearly constant. Although carbon-14 is removed from the atmosphere by biological processes, new carbon-14 is produced by the interaction between cosmic rays and atmospheric nitrogen.) A bone or plant sample that is one half-life old—that is, 5,730 years old—contains only half as much carbon-14 as a living organism. Figure 11.21 shows part of a scroll from a collection known as the Dead Sea Scrolls, discovered in Israel in 1947 and dated to 125 BC by carbon-14 dating. Carbon-14 dating can be used on samples back to about 60,000 years old; anything older than this contains too little carbon-14 to be measured accurately.

Learning Check 11.4

1. Compare and contrast relative-age dating and absolute-age dating.
2. What is the difference between a stable isotope and an unstable isotope?
3. Give two reasons why uranium-lead dating cannot be used to date human-made artifacts, such as the Dead Sea Scroll shown in Figure 11.21.
4. Why can't carbon-14 dating be used to date rocks such as granite?

11.5 Sedimentary Environments

Sedimentary rocks are formed in distinctive *depositional environments*, such as deltas, shorelines, reefs, river floodplains, and deserts. A depositional environment is a type of place with distinctive characteristics that results in deposition of particular types of sediment. We can speak of a beach depositional environment or a glacial depositional environment, as well as other places where sediments form, as illustrated in Figure 11.22. Rock layers do not have signs on them saying things like "formed on a beach," so geologists have to do a considerable amount of detec-

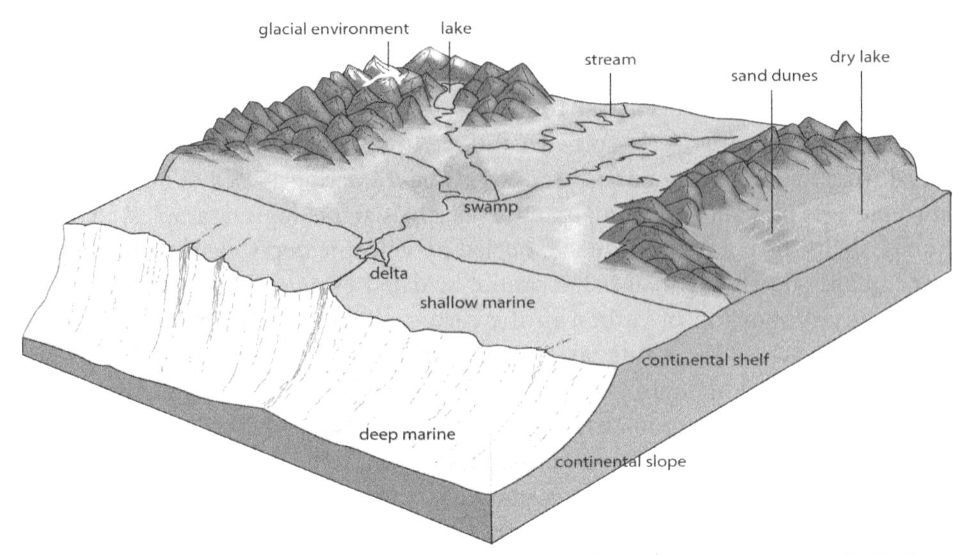

Figure 11.22. Sediments are deposited in a wide variety of depositional environments, each of which has distinctive types of sediments and fossils.

tive work in order to determine the type of depositional environment in which any given sedimentary rock layer formed.

An important part of understanding how sedimentary rocks form is the study of modern depositional environments. Layers of sand, for example, form in a number of different settings such as beaches, deltas, meandering stream point bars, and sandy deserts. Sand deposited in each of these environments has certain characteristics. For example, layers of sand deposited in a delta are generally underlain or overlain by layers of silt or clay, which are deposited outside the main stream channel, while sand dunes in a desert rarely have much clay or silt because the fine-grained sediments have been carried away by wind. Another important clue for determining depositional environment is the types of fossils and trace fossils present. Sand deposited in a shallow ocean contain marine fossils, while sand deposited on land might have bones and tracks of land animals. When studying an

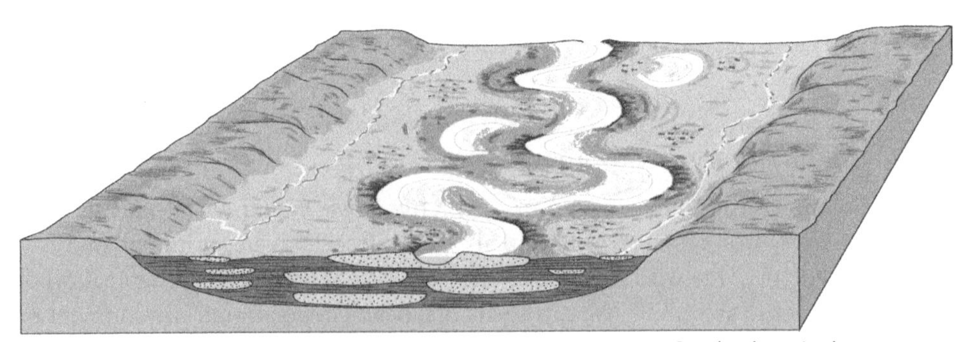

Figure 11.23. A modern meandering stream depositional environment. Sand is deposited near stream channels, and silt and clay are deposited in the floodplain. The result is a deposit that contains scattered layers of sand surrounded by silt and clay. The deeper layers of sand were formed when the channel was at different positions on the floodplain.

ancient sandstone layer, geologists compare the ancient rock to sediments formed in modern environments and try to determine which depositional environment is the best match. This is an application of the principle of uniformitarianism, which is sometimes expressed as "the present is the key to the past."

Let's look at some typical depositional environments. For each of these environments, we will look at the types of sediment deposited in modern settings, and at an ancient example from the rock record.

Meandering Streams

The two most prominent features along a meandering stream are the stream channel and the floodplain. Water in the stream channel moves rapidly enough to keep silt and clay in suspension, but sand will settle to the bottom in places where the current moves more slowly. Because of this, stream-channel deposits are composed largely of sand, much of which is deposited on point bars. When a stream is at flood stage, however, the water spreading out over the floodplain moves more slowly, and fine-grained sediment grains are able to settle out of the water, leaving layers of silt and clay. The result is a deposit that contains sandy parts surrounded by layers of silt and clay, as illustrated in Figure 11.23. Figure 11.24 shows a sandstone deposit that likely formed in a meandering stream depositional environment.

Figure 11.24. This sandstone outcrop is interpreted to have been part of a point bar along a meandering river.

Deltas

In some ways, sedimentary deposits created by deltas are similar to deposits created by streams. Like meandering streams on floodplains, deltas have stream channels where sand is deposited, and marshy areas between stream channels where silt and clay are deposited, along with organic material. In general, deltas

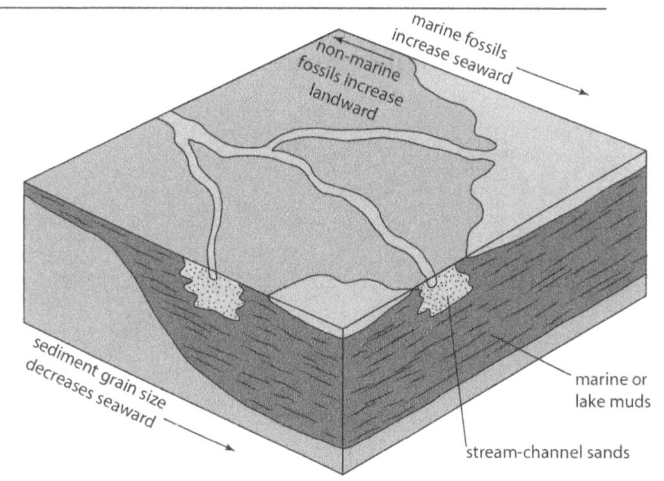

Figure 11.25. A modern delta environment, such as the one at the mouth of the Mississippi River.

Figure 11.26. The inclined layers near the center of this image taken on the surface of Mars are interpreted to have formed in an ancient delta. Water would have been flowing from right to left.

have coarser sediments on their landward sides and finer sediments deposited in deeper water, as shown in Figure 11.25. One thing that sets delta deposits apart from stream deposits is the presence of marine fossils (or lake fossils if the delta formed in a lake rather than the ocean).

Delta-like deposits are common in the sedimentary rock record on Earth. Figure 11.26 shows sedimentary rocks on the surface of Mars that have been interpreted to have formed in an ancient delta.

Lakes

Sediments deposited in lakes vary from place to place depending on what part of a lake the sediments were deposited in. Typically, coarser sediments such as sand are deposited near the shoreline, and finer sediments such as clay and silt are deposited in deeper water at some distance from the shore. A key characteristic, however, for recognizing ancient lake deposits, is the fossils contained in the rocks. Figure 11.27 shows a fossil freshwater fish from the Eocene Green River Formation found in Wyoming, Colorado, and Utah. The rocks of the Green River Formation have zones, with near-shore deposits containing shoreline plants (cattails, lily pads), freshwater snails, wading birds, turtles, frogs, and crocodiles around the margins of the formation, and fine-grained deeper water deposits in the central areas of the formation.

Figure 11.27. A freshwater fish fossil from the Green River Formation, Wyoming.

Wind-Deposited Sand Dunes

A distinguishing characteristic of wind-blown sand dunes is that they are made almost entirely of sand, with almost no silt, clay, or gravel mixed in. In windy places, silt and clay are carried away by wind in dust clouds. Sediment grains larger than sand, on the other hand, are too massive to be carried by wind, and are not deposited in wind-blown deposits. Another characteristic of sand-dune deposits is the presence of large scale *cross-bedding*, illustrated in Figure 11.28. Cross-beds are thin sloping layers within thicker layers, formed by dunes or ripples in sediments deposited either by wind or water. Each tilted layer in Figure 11.28 represents a former dune slip face. There are different types of cross-bedding in sediments, but wind-blown cross-beds are much larger than most other types of cross-bedding. A third characteristic of sand dune deposits is the absence of marine fossils. Sand dunes formed in deserts might have trace fossils formed by lizards and other vertebrates—and even insects and spiders—but do not contain body fossils of marine organisms such as marine clams or fish. Fossils are critical for distinguishing ancient wind-blown sand deposits from water-deposited sand deposits.

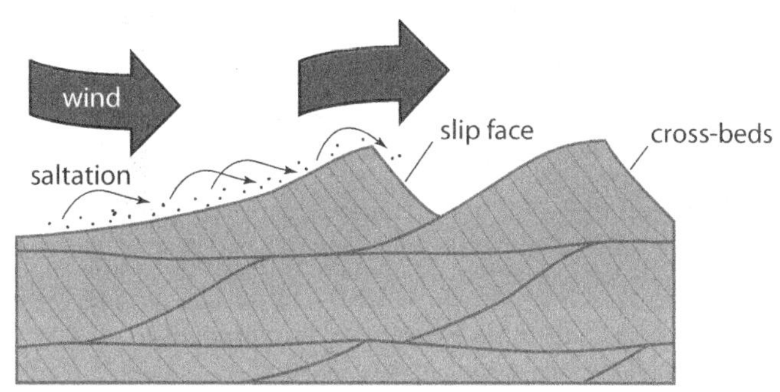

Figure 11.28. Cross-bedding forms in modern sand dunes as sand falls down the slip face as dunes migrate downwind.

Figure 11.29 shows an ancient wind-blown sand deposit, the Navajo Sandstone of the southwestern United States. Like modern wind-blown sand deposits, the Navajo is composed mostly of pure sand, has large scale cross-bedding, and has tracks of terrestrial animals, such as dinosaurs and extinct mammals.

Figure 11.29. The Jurassic Navajo Sandstone of the southwestern United States formed in an ancient sandy desert.

Reefs

Figure 11.30. Coral reefs are built by small coral animals living in colonies numbering in the millions of individuals.

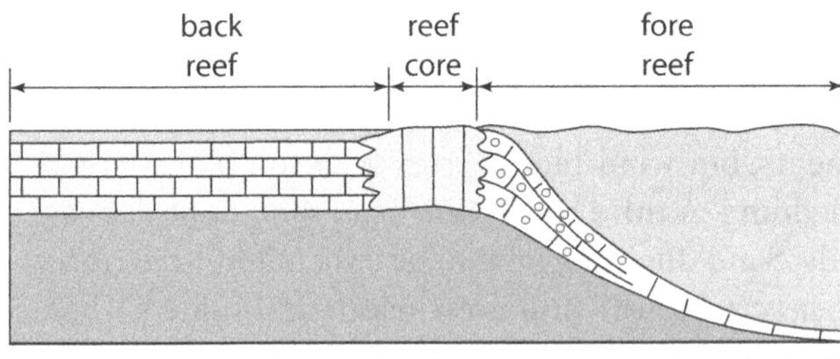

Figure 11.31. The reef depositional environment consists of a back reef, reef core, and fore reef. The fore reef faces the open ocean, and the back reef is protected from large waves by the reef core.

Coral reefs form in shallow, tropical seas. The reefs are produced by coral animals, shown in Figure 11.30. These organisms are basically mouths surrounded by tentacles which they use to gather food that floats by. Despite their lack of mobility,[5] corals are not plants; they must eat in order to survive. Corals produce a calcium carbonate foundation to which they stay attached throughout their adult lives. Most corals live in colonies composed of many thousands or millions of individuals and build up massive structures known as reefs.

Figure 11.31 is a diagram of a typical coral reef with three parts, each composed of a different type of limestone. The fore reef faces the open ocean and takes the brunt of the force of incom-

Figure 11.32. The rocks composing the sheer cliffs of the Guadalupe Mountains in Texas were formed in ancient reefs.

5 Coral larvae can swim and look like tiny jellyfish; coral adults attach to the sea floor

ing waves. The fore reef is usually steep and contains blocks that have broken off from higher up and rolled down the slope. The reef core is in shallow water and may even be partially exposed to the air at low tide. The limestone formed in the reef core contains numerous fossils that grew in place. The back reef is on the landward side of the reef. This area usually consists of a shallow body of water called a lagoon. The limestone that forms in the back reef lagoon is generally fine-grained limestone, composed of microscopic fragments of hard parts of organisms.

Some ancient reefs are composed of fossil corals, just like modern reefs. At various times in Earth history, other organisms, such as algae, clam-like organisms, and sponge-like organisms, have been important reef-builders. Figure 11.32 shows an ancient reef exposed in the Guadalupe Mountains of western Texas. This ancient reef was formed by sponges and algae rather than by corals, but still has rocks representing the back reef, reef core, and fore reef, just like modern reef depositional environments.

Learning Check 11.5

1. How might fossils help a geologist determine whether or not a layer of shale was deposited in a river floodplain, as part of a river delta on a coastline, or in a deep marine environment? (It will be helpful to consider the types of organisms that live in each of those environments.)
2. Large sand dunes can also form under water, in areas with strong bottom currents. How might these be distinguished from wind-blown sand dune deposits?

11.6 An Overview of Earth History

By using absolute-age dating methods such as uranium-lead dating, geologists now believe Earth is about 4.6 billion years old. Earth's surface has constantly changed as tectonic plates move, mountain ranges form, and the ranges erode down to flat plains. There has been some form of life on Earth for at least 3.5 billion years; perhaps longer. Throughout Earth history, different types of organisms, ranging from microscopic algae to dinosaurs, have flourished and then gone extinct. The 6,000 years of written human history make up just a blink at the end of Earth history.

11.6.1 Divisions of Earth History

Throughout the late 1700s and early 1800s, scientists gained a basic understanding of Earth history by applying the principles of relative-age dating. One key to this effort was the realization that fossils occur in a definite order from older rocks to younger rocks. Earth's oldest sedimentary rocks contain only fossils of the simplest organisms, such as bacteria and other single-celled organisms. Slightly younger rocks contain fossils of shelled marine invertebrate organisms. The next

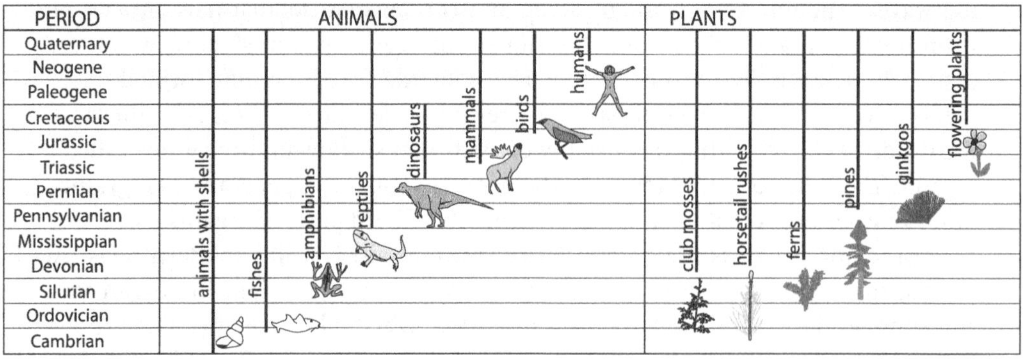

PERIOD	ANIMALS	PLANTS
Quaternary		
Neogene		
Paleogene		
Cretaceous		
Jurassic		
Triassic		
Permian		
Pennsylvanian		
Mississippian		
Devonian		
Silurian		
Ordovician		
Cambrian		

animals with shells, fishes, amphibians, reptiles, dinosaurs, mammals, birds, humans

club mosses, horsetail rushes, ferns, pines, ginkgos, flowering plants

Figure 11.33. The principle of fossil succession states that fossils appear in a certain order in the fossil record.

group of fossils to appear in the fossil record is the fishes, followed by amphibians, reptiles, and finally, mammals. The order in which fossils occur in the fossil record is known as the *principle of fossil succession*, illustrated in Figure 11.33.

The fossil record is actually much more complex than the figure indicates. For example, the reptiles found in the oldest reptile-bearing rocks ("Pennsylvanian") are very different from the dinosaurs and other reptiles that lived in rocks we know as "Jurassic." And the Jurassic reptiles are for the most part very different from the reptiles alive today.

The *geologic column* (also called the *geologic time scale*) is an ordered arrangement of rock layers, from oldest to youngest, that geologists have pieced together from the rock layers found in different places on Earth. The geologic column represents a time line of Earth history. As geologists recognized fossil patterns, they started to give names to time periods that contained distinctive collections of fossils.

Organisms found near the top of the geologic column are similar to organisms found on Earth today, and most of the organisms lower down in the geologic column are quite different from anything alive at present. Beginning in the mid-1900s, scientists were able to use radiometric dating to put absolute ages on the geologic column. The geologic column, shown in Table 11.3, is divided into time units called eons, eras, periods, and epochs. You will find it helpful to refer to Table 11.3 often as you continue reading this section.

The longest division on the geologic time scale is the *eon*. There are four eons, each of which lasted hundreds of millions of years. The four eons are the Hadean, Archean, Proterozoic, and Phanerozoic eons. The Hadean, Archean, and Proterozoic eons are often grouped together as Precambrian time. It is difficult for us to imagine time periods lasting hundreds of millions or even billions of years.

Eons are divided into *eras*. The Phanerozoic Eon, for example, is divided into three eras: the Paleozoic, Mesozoic, and Cenozoic. Paleozoic means "old life," Mesozoic means "middle life," and Cenozoic means "recent life." The eras of the Phanerozoic and Proterozoic eons are further divided into *periods*, some of which are shown in Table 11.3. Several of the periods were named after the location where rocks of that period were first studied. Devonian rocks were first studied in Dev-

Eon	Era	Period		Epoch	Time (Ma)	Geologic History
Phanerozoic	Cenozoic	Quaternary		Holocene	0.01–2.6	Ice ages. Modern plants and animals.
				Pleistocene	2.6–0.01	
		Neogene		Pliocene	5.3–2.6	Mammals and flowering plants dominate on land. Uplift of Alps and Himalayas.
				Miocene	23–5.3	
		Paleogene		Oligocene	34–23	
				Eocene	56–34	
				Paleocene	66–56	
	Mesozoic	Cretaceous			145–66	Dinosaurs continue to dominate on land. Rocky Mountains begin to uplift. Mass extinction at end of period.
		Jurassic			201–145	Dinosaurs dominate on land. First birds appear.
		Triassic			252–201	First mammals appear. Atlantic Ocean begins to open (breakup of Pangea).
	Paleozoic	Permian			299–252	Building of Appalachians ends. Mass extinction at end of period.
		Carboniferous	Pennsylvanian Period		323–299	Coal swamps. Amphibians dominate on land.
			Mississippian Period		359–323	
		Devonian			419–359	Appalachians continue to form.
		Silurian			444–419	First land plants appear; insects.
		Ordovician			485–444	First fish. Appalachians begin to form.
		Cambrian			541–485	Abundant invertebrates.
(Precambrian) Proterozoic					2500–541	Bacteria-like organisms dominate. Several periods of mountain-building followed by intense erosion.
Archean					4000–2500	Few evidences of life. Formation of core of continents.
Hadean					4600–4000	Few surviving rocks or minerals. No fossils.

Notes

1. "Precambrian" is not an official division of geologic time, but is commonly used to group the Hadean, Archean, and Proterozoic eons together.
2. The Mississippian and Pennsylvanian periods are recognized in North America; in the rest of the world that time is lumped together as the Carboniferous Period.
3. Older geologic time scales group the Paleogene and Neogene periods together as the Tertiary Period.

Table 11.3. Geologic Time Scale.

onshire, England and Jurassic rocks were first studied in the Jura Mountains of southeastern France. An *epoch* is a division of a period. Even though epochs are the shortest time unit we will consider in this book, an epoch is still a vast period of time—millions of years.

These divisions of geologic time are defined by the types of fossils found in layers of sedimentary rocks. There are substantial differences between fossils found in different eons and eras; the differences become less dramatic between consecutive periods or epochs. Rock layers are always found in the order of the geologic column, unless the layers have been disturbed by folding or faulting. From bottom to top, layers are found in the order Cambrian–Ordovician–Silurian–Devonian, and so on (but never in a different order, such as Jurassic–Ordovician–Permian–Cambrian). In most locations, there are layers missing from the complete geologic column, but there are places where there are layers from each period of the Phanerozoic Eon, from Cambrian up to Quaternary.

11.6.2 The Precambrian

Imagine a world where the land surface has no plants or animals, only barren rock. The oceans on this world have no fish, whales, jellyfish, lobsters, corals, clams, or any of the many thousands of creatures we see in today's oceans, only green, slimy masses made up of billions of microscopic cells of bacteria. This is what Earth was like for most of its history, the time designated as Precambrian in Table 11.3.

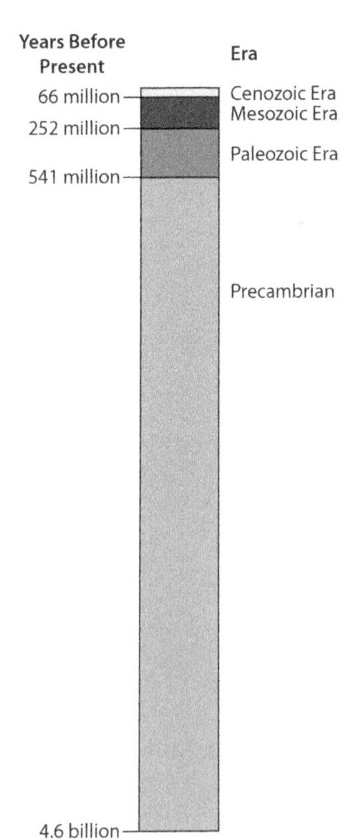

Figure 11.34. Most of Earth history occurred in Precambrian. The Cenozoic, which is the present era, is the shortest era of geologic time.

The Precambrian covers a vast extent of time, from 4.6 billion years ago up until 541 million years ago—88% of Earth history, as illustrated in Figure 11.34. The earliest eon of the Precambrian is the Hadean. Our present understanding is that Earth was a very different place during the Hadean on. At the beginning of the Hadean, Earth may have been largely molten. Over time, this magma would have crystallized to form Earth's earliest crust. Later in the Hadean, Earth had oceans, but Earth's atmosphere would have been low in oxygen and toxic to most life forms. We have no surviving complete rocks from the Hadean, but geologists have discovered individual crystals of a mineral called zircon which they believe were formed in the Hadean. These near-microscopic crystals have been dated at 4.4 billion years.

The oldest complete rocks in Earth's crust are igneous and metamorphic rocks from the Archean Eon, such as the gneiss shown in Figure 11.35, dated at 3.7 billion years. Large portions of the central areas of continents are underlain by rocks of Archean age. The only fossils preserved from the Archean Eon were formed by microscopic bacte-

Figure 11.35. These Archean metamorphic rocks in Greenland have been dated at about 3.7 billion years old, which make them among the oldest rocks exposed in Earth's crust.

Figure 11.36. These Proterozoic sedimentary rocks in Glacier National Park have ripple marks and mud cracks, indicating that the rocks formed in a near-shore environment that occasionally dried out.

ria. Many of the world's most valuable mineral deposits are found in Archean rocks.

The Proterozoic Eon lasted from 2.5 billion years ago to the beginning of the Cambrian Period 541 million years ago. Many sedimentary rocks formed during the Proterozoic Eon are not as highly metamorphosed as Archean rocks, so it is easier for scientists to reconstruct what Earth was like in the Proterozoic. Figure 11.36 shows Proterozoic rocks with ripple marks like those formed on beaches and mud cracks like those formed when mud starts to dry out, indicating that these rocks formed in a shallow-water environment.

Figure 11.37. The layers in stromatolites were formed by thin mats of photosynthetic bacteria.

The only fossils preserved from throughout most of the Proterozoic Eon are bacteria. These bacteria formed slimy mats or layers that trapped sediment grains. These are preserved as structures called stromatolites, shown in Figure 11.37. These bacteria were of a type that used photosynthesis, just as plants do. In photosynthesis, organisms take in carbon dioxide from the atmosphere and produce oxygen. It is believed that photosynthesis was responsible for changing Earth's atmosphere from being oxygen-poor in the Archean to oxygen-rich as it is at present. The presence of oxygen allowed Earth to become the home to more complex organisms by the end of the Proterozoic, such as animals similar to marine worms and jellyfish.

11.6.3 The Paleozoic Era

In sharp contrast to the Proterozoic Eon and earlier, rocks from the Paleozoic Era have an abundant fossil record. Rocks deposited at the end of the Proterozoic Eon contain occasional fossils of a few species of soft-bodied organisms but Cambrian rocks contain fossils from a great diversity of invertebrate life forms. This tremendous increase in the variety of organisms on Earth is known as the *Cambrian explosion*. The Cambrian explosion is one of the most fascinating events in the history of life on Earth. It is as if a marine aquarium had only algae growing on the glass, and was quickly transformed to include hundreds of different types of creatures. Cambrian rocks include fossils of organisms such as sponges, snails, corals, and brachiopods—a group of shelled organisms that look somewhat like clams. One very common fossil type from Cambrian rocks is the trilobites, such as the one shown in Figure 11.38. Trilobites were arthropods, which are invertebrates with jointed legs such as crabs and lobsters (in the sea) and insects and spiders (on land). The most striking difference between Cambrian animals and earlier organisms is the presence of hard shells. The hard shells make organisms much more likely to be preserved as fossils.

Figure 11.38. Trilobites were abundant in the Cambrian Period and lived until the end of the Paleozoic Era. The fossil shown is large—about 12 inches long. Most trilobites range from 1 to 4 inches in length.

Each period in the Paleozoic Era had a distinct collection of organisms—life forms now preserved in the fossil record. For instance, the first appearance of fish in the fossil record is in the Ordovician Period, but these fish were very different from modern fish. Ostracoderms, illustrated in Figure 11.39, were jawless fish with bony plates that covered their heads. The Silurian

Figure 11.39. Artist's rendering of ostracoderms, a type of fish that first appears in Ordovician Period rocks. They became extinct by the end of the Devonian Period.

Figure 11.40. Dimetrodons lived in the Permian Period and were in some ways more like mammals than like dinosaurs.

oceans had giant scorpion-like creatures called eurypterids, some of which grew to be considerably larger than a human. The first land animals appear in Silurian layers and Devonian rocks contain fossils of the first true amphibians. The amphibians dominated the land in the Carboniferous Period. The Carboniferous Period had abundant forests and swamps that contained giant insects, such as dragonflies with wingspans of 70 centimeters. By the Permian Period, the largest land animals were the synapsids, a group that people often confuse with dinosaurs, and which included animals such as Dimetrodon, illustrated in Figure 11.40.

Several times during the Paleozoic Era, North America was almost completely covered by shallow seas. Evidence for this includes the presence of marine sediments and marine fossils spread across the continent. Similar conditions existed on other continents at various times during the Paleozoic. At other times in the Paleozoic, significant thicknesses of terrestrial sedimentary rocks were deposited.

In the discussion of plate tectonics in Chapter 6, we saw that the Appalachian Mountains in the United States were formed by a collision between Africa and North America. Geologists have reconstructed the position of the continents throughout the Phanerozoic Eon, illustrated by the map sequence shown in Figure 11.41. In the Cambrian Period, the continents were scattered across much of Earth's surface. Throughout the Paleozoic, the continents slowly moved closer to each other at a rate of a few centimeters per year. By the Permian Period—the last period of the Paleozoic Era—the continents had assembled to form one supercontinent, Pangea.

The Permian Period ended with a *mass extinction* event in which 90% of the marine fossil species and many land species were wiped out. A mass extinction occurs when many species go extinct at about the same time. Types of organisms that became extinct at the end of the Permian include many species of corals, brachiopods, and insects, as well as all of the trilobite species. Geologists are not certain as to the cause of this extinction. But whatever happened, it affected life around the globe.

11.6.4 The Mesozoic Era

During the Mesozoic Era, Pangea began to break apart and the continents began to creep toward their present positions. The Atlantic Ocean started to open in the Jurassic Period, as shown in Figure 11.41D. As the Atlantic Ocean was widening, the western edge of North America became part of a convergent plate boundary, with subduction of oceanic crust. This led to several episodes of mountain-

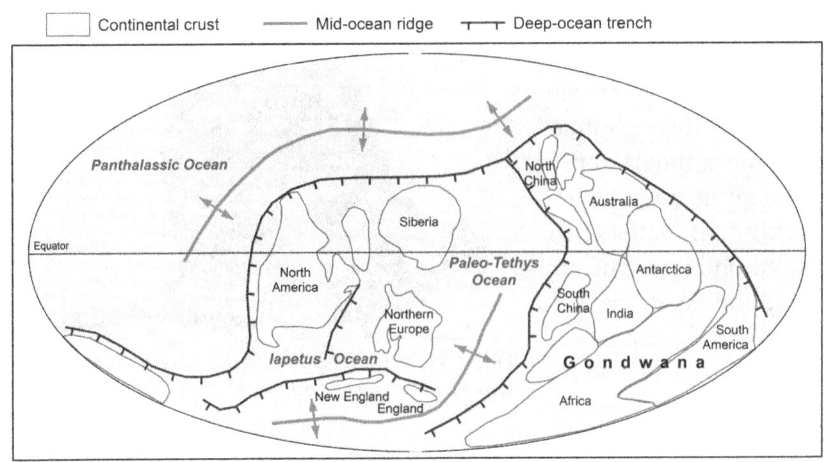

Figure 11.41A. Inferred positions of the continents in the Ordovician Period, about 460 Ma (million years ago). Throughout much of the Paleozoic Era, the southern continents were connected together to form a large continent called Gondwana.

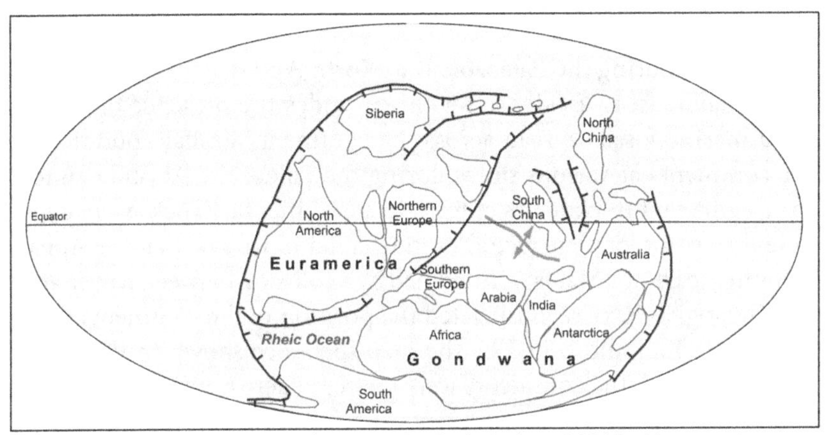

Figure 11.41B. Positions of the continents in the Devonian Period, about 390 Ma. Most continents were slowly moving closer to one another.

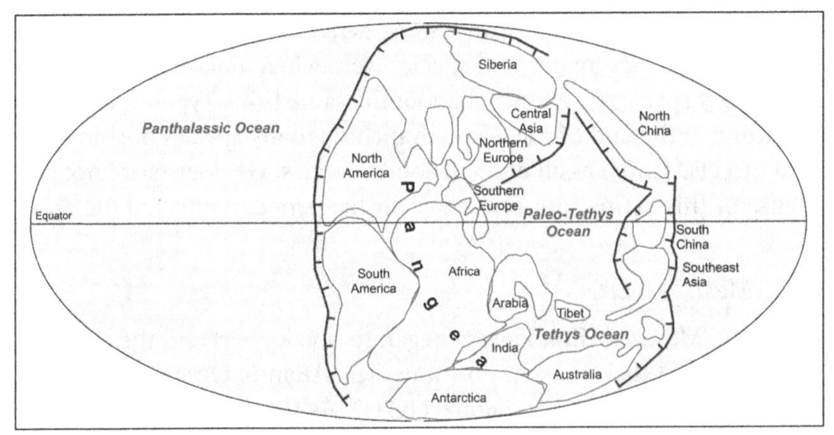

Figure 11.41C. Positions of the continents during the Permian, about 250 Ma, when most continents were combined into one supercontinent, Pangea.

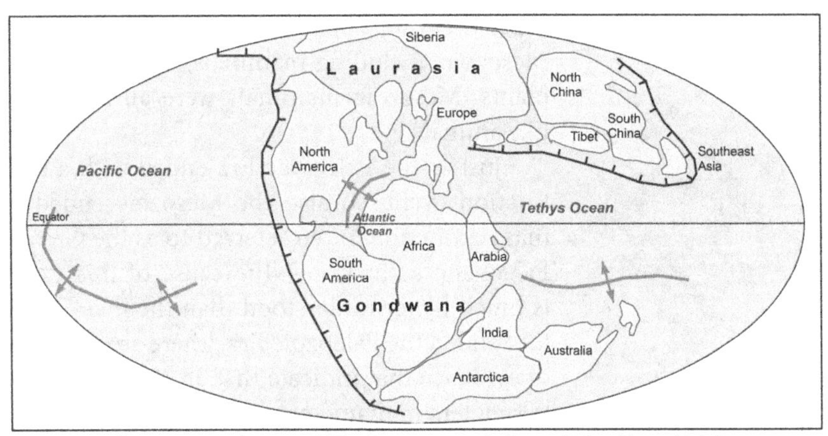

Figure 11.41D. During the Jurassic, about 150 Ma, Pangea was being split by rifts, which became oceans such as the Atlantic.

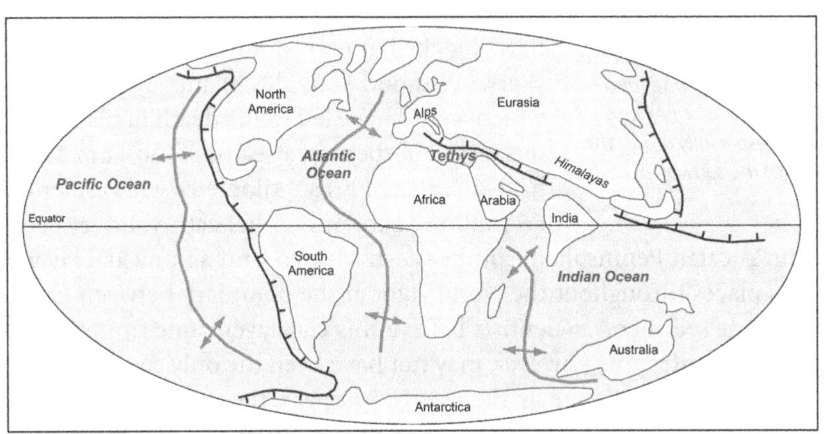

Figure 11.41E. By the Eocene Epoch, about 50 Ma, the continents were moving closer to their present positions. The Himalayas and Alps were beginning to form as India and Africa started to collide with Eurasia.

building in the western part of the continent, complete with extensive volcanism, the intrusion of the large batholiths (such as shown in Figure 7.34), and thrust faulting (such as shown in Figure 6.33). In the Triassic and early Jurassic periods, large areas of western North America were covered by desert sand dunes, leading to the formation of sandstone layers such as the Navajo Formation shown in Figure 11.29. Later, in the Cretaceous Period, sea level rose, forming an inland sea that extended from the modern Gulf of Mexico all the way up to the Arctic Ocean. This inland sea, called the Western Interior Seaway, is illustrated in Figure 11.42.

The Mesozoic Era is often called the Age of Dinosaurs. Paleontologists have identified over 700 species of dinosaurs from the Mesozoic, ranging from the size of a turkey up to gigantic types 40 meters (130 feet) long from head to tail, such as the Apatosaurus shown in Figure 11.13. The Mesozoic Era also had large marine reptiles, such as those illustrated in Figure 11.43, and flying reptiles that had greater wingspans than any bird that has ever lived. These swimming and flying reptiles are not considered to be dinosaurs. Several other important groups of organisms

Figure 11.42. During much of the Cretaceous Period, a shallow sea extended from the Gulf of Mexico up to the Arctic Ocean. Marine fossils are found within this seaway, and land fossils, such as dinosaur bones and tracks, are found especially along the western border of the seaway.

made their first appearance in the fossil record in the Mesozoic, including mammals, birds, and flowering plants. Mesozoic mammals were all small, rodent-like animals.

Just as the Paleozoic Era ended with a mass extinction event, so also the Mesozoic ended with a mass extinction, often referred to as the Cretaceous-Paleogene extinction. The cause of this extinction is much better understood than the mass extinction that ended the Paleozoic Era. There are multiple lines of evidence that indicate that an asteroid perhaps 10 kilometers in diameter crashed into Earth at the end of the Cretaceous Period. This impact would have had global consequences, including throwing massive clouds of dust into the atmosphere that would have blocked much of the sun's light for perhaps years. Without sunlight, plants would have died, which would have led to the death of plant-eating dinosaurs, and then meat-eating dinosaurs. Geologists have discovered a 180-kilometer wide crater dated at 66 million years buried beneath younger sediments under the Yucatan Peninsula in southeastern Mexico, and an unusual layer of clay at various places throughout the world right at the boundary between Cretaceous and Paleogene sediments. Scientists believe this clay layer came from this asteroid impact. This catastrophe, however, may not have been the only cause of the extinction event. At the same time as the asteroid impact, there were massive volcanic eruptions in India—with a volume of a half-million cubic kilometers of lava—that also would have had global effects.

The Cretaceous-Paleogene extinction event wiped out the dinosaurs as well as some important groups of marine organisms. However, mammals survived this extinction and flourished in the Cenozoic Era.

11.6.5 The Cenozoic Era

The world of the Precambrian, Paleozoic, and Mesozoic eras would have seemed strange to us. Not only would a map of Earth look very different, with continents all in the wrong places, but the world was populated by all sorts of organisms that were quite different from those we are familiar with. But in the Cenozoic Era things started to become more like what we recognize. As indicated by the final map in Figure 11.41, during the Cenozoic Era the Atlantic Ocean continued to grow wider and the continents drifted towards their present arrangement. We have a much more complete record of the Cenozoic than we do for earlier times in Earth history, so geologists usually refer to the epochs—such as Eocene and Miocene—rather than to the periods of the Cenozoic.

Figure 11.43. The largest predators in Mesozoic seas were marine reptiles of a variety of types, such as the Augustasaurus, *here artistically depicted in its habitat.*

Earth's climate was up to 10°C warmer during the Paleocene and Eocene Epochs than it is today. Fossil evidence indicates that even polar areas had dense forests, indicating a climate that was warm throughout the year. By the Oligocene Epoch, in the middle of the Cenozoic Era, Earth's climate had begun to cool down, and the forests that once grew on Antarctica were replaced by glaciers. This cooling trend continued through the Pleistocene Epoch, during which thick ice sheets covered large parts of North America and Eurasia. The Holocene Epoch in which all of human history has occurred may be considered to be just a brief warm spell during the Pleistocene Ice Age.

The Cenozoic Era is known as the Age of Mammals. The mammals living at the beginning of the Paleocene Epoch were mostly small organisms similar to mice or shrews. The fossil record shows that both the body size and variety of mammal types increased throughout the Paleocene and Eocene. In other words, mammals got larger, and there were more types of mammals. With each epoch, the types of organisms that lived on Earth—not just mammals but all types of plants and animals—became more and more similar to organisms that live on Earth today. By the Pleistocene Epoch, organisms present on Earth looked very similar to organisms that are alive today, as illustrated in Figure 10.41.

The Holocene Epoch is the Age of Humans, who have been created in the image of God. As discussed in Chapter 3, God has placed humans in a position of responsible stewardship over the rest of creation. With our great population and technology, humans have the capability to be wise stewards of the creation. Because of our sin, humans also have the capacity to do great harm to the rest of the creation. The choice is ours.

Learning Check 11.6

1. State the principle of fossil succession.
2. Distinguish between eons, eras, periods, and epochs.
3. Describe what Earth was like during the Precambrian.
4. Describe what Earth was like during the Paleozoic Era.
5. Describe what Earth was like during the Mesozoic Era.
6. Describe what Earth was like during the Cenozoic Era.

Chapter 11 Exercises

Answer each of the questions below as completely as you can. Write your responses in complete sentences unless instructed otherwise.

1. In what ways is the geological principle of uniformitarianism compatible with Christianity?
2. Some Christians believe Earth is billions of years old and other Christians believe Earth is only a few thousand years old. How can Christians cooperate lovingly as they consider matters such as this?
3. In diagram A, there are pieces of granite gravel enclosed in sandstone. In diagram B, there are pieces of sandstone enclosed in granite. For each diagram, make a hypothesis for how the pieces of one rock type ended up enclosed in the other. One of these situations was depicted in Figure 7.32.

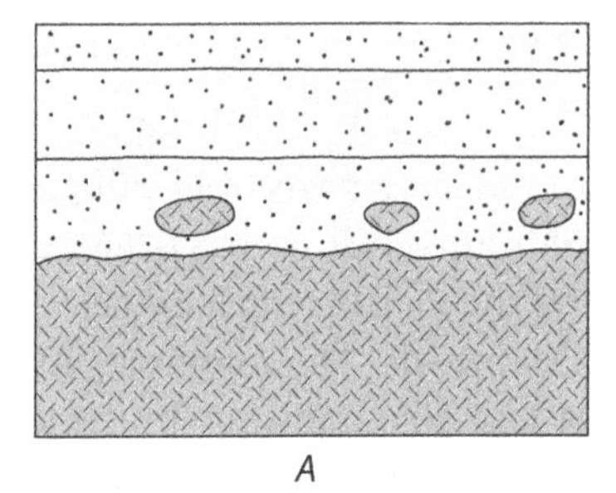

A

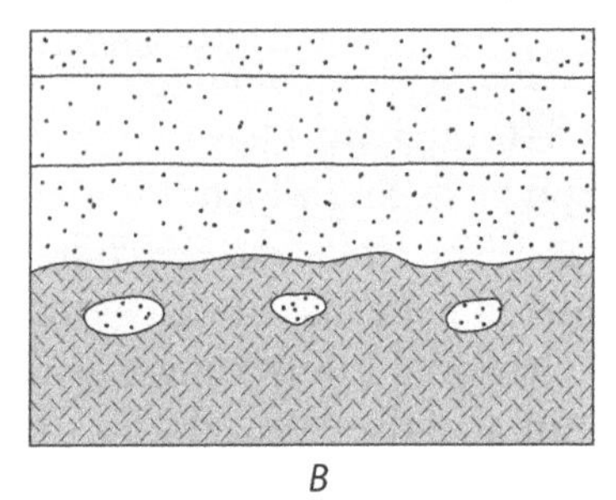

B

4. Use the diagram at the right to answer the following questions.
   a. Which is the oldest rock unit?
   b. Which is the youngest rock unit?
   c. Which rock units were formed before faulting? Which rock units were formed after faulting?
   d. Unconformities exists between which rock layers?
   e. Did the folding of rock layers occur before or after the erosion that caused the unconformity? Explain.
   f. What would have happened to the sedimentary rock layers that came into contact with the igneous intrusion (see Section 5.4.1).
   g. List the order of geologic events represented by the diagram at the right.
5. Fill in the blank boxes in the table of nuclides at the top of the next page. (Refer to the periodic table in Figure 4.3.)

Nuclide	Protons	Neutrons
oxygen-16		
oxygen-17		
oxygen-18		
lead-204		
lead-206		
lead-207		
plutonium-244		
	11	12
	18	22
	20	20
	26	30
	79	118

6. Complete the following table, showing the fraction of the parent isotope remaining during radioactive decay from zero to ten half lives.

Number of Half-Lives	Fraction of Parent Remaining	Number of Half-Lives	Fraction of Parent Remaining
0	1	6	
1	1/2	7	
2		8	
3		9	
4		10	
5			

7. Based on the fractions in the table for question 6, if a bone sample in an archeological site has only 1/32 of its original carbon-14 remaining, about how old is it?
8. If a mineral in granite has equal amounts of uranium-235 and lead-207, how old is it? Assume that the mineral did not have any lead in it when it formed.
9. If an igneous rock was formed in the Cretaceous Period, would you expect its minerals to have more lead-206 than uranium-238, less lead-206 than uranium-238, or about the same of each? Again, assume that the rock contained no lead when it crystallized.
10. According to the principle of fossil succession, would you expect to find fossils of humans in the same sedimentary rock layers as dinosaur fossils? Why or why not?
11. Why do you think the Phanerozoic Eon is divided into eras, periods, and even epochs, but the Hadean Eon is not?

Experimental Investigation 8: Fossil Identification

Note:

The text for this experiment was not complete at the time this volume went to press. For those using the Resource CD, the experiment may be found in the share folder online. Those using this text but not the CD may obtain a pdf file of the experiment from Novare Science & Math by sending an email to info@novarescienceandmath.com. Simply request Earth Science Experiment 8 and we will send it to you.

Chapter 12

Oceanography

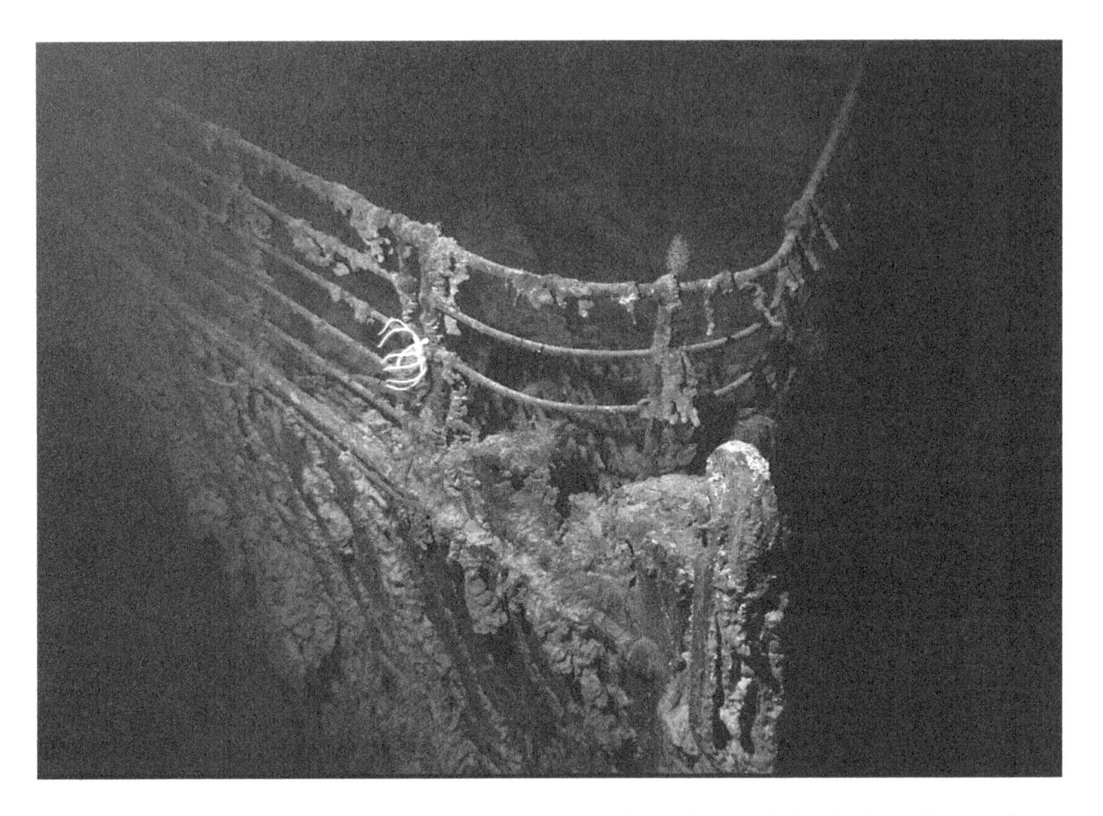

On April 15, 1912, the luxury passenger liner RMS Titanic *struck an iceberg in the North Atlantic Ocean and sank in deep water, with a death toll of more than 1,500 people. The wreck of the* Titanic, *pictured above, remains on the seafloor 600 kilometers off the coast of Newfoundland, Canada, at a depth of about 3,800 meters.*

It has always been difficult to explore the seas, and even more difficult to explore the depths of the seas. The Titanic *is a large vessel, but its wreckage on the seafloor is tiny compared to the vast immensity of the ocean. Even though people knew roughly where the Titanic was when she sank, attempts to find the ship using echo sounding drilling ships and other methods all failed. Researchers were not able to find the wreck of the Titanic until 1985, when remote-controlled submersible vessels were used to take pictures such as this image of the bow of the ship.*

Before the wreck was discovered, many people thought that the ship would still be in good enough condition to be raised to the surface. They assumed that the cold, dark waters at the bottom of the ocean would preserve the ship in almost perfect condition. The expeditions that explored the wreck of the Titanic, *however, discovered that not only had the ship broken into two large pieces as she sank, but that further damage had been done when the ship struck the seafloor. In addition, metal-eating microbes are consuming the iron hull of the ship, creating orange icicle-like formations such as those visible in the photograph. These microbes will eventually reduce the wreck of the* Titanic *to a pile of rust on the seafloor.*

In this chapter, we will explore the mysterious oceans. Some of the topics we will address include the properties of seawater, ocean currents, waves, tides, and shorelines. We will also consider the various oceanic environments and the types of organisms that inhabit them, as well as human interactions with the oceans.

Objectives

After studying this chapter and completing the exercises, you should be able to do each of the following tasks, using supporting terms and principles as necessary.

1. Describe the distribution of oceans and seas on Earth's surface.
2. Describe the chemical composition and physical properties of seawater.
3. Explain how and why temperature changes with depth in the oceans.
4. Describe the major ocean surface currents and explain how these currents form.
5. Describe the properties of waves.
6. Explain how tides form, and describe typical tidal rhythms.
7. Describe ocean environments as they relate to life.
8. Describe the major types of erosional and depositional shoreline landforms.

Vocabulary Terms

You should be able to define or describe each of these terms in a complete sentence or paragraph.

1. abyssal zone
2. aquaculture
3. barrier island
4. bathyal zone
5. bay mouth bar
6. beach
7. benthic
8. benthos
9. brackish
10. breaker
11. crest
12. current
13. groin
14. gyre
15. hadal zone
16. intertidal zone
17. lagoon
18. longshore current
19. mangrove
20. neap tide
21. nekton
22. neritic zone
23. nutrient
24. oceanic zone
25. oceanographer
26. oceanography
27. pelagic
28. period
29. photic zone
30. phytoplankton
31. plankton
32. salinity
33. sea arch
34. sea ice
35. sea stack
36. spit
37. spring tide
38. sublittoral zone
39. thermocline
40. tidal bulge
41. tidal flat
42. tidal range
43. tide
44. trough
45. undertow
46. upwelling
47. wave
48. wave height
49. wave-cut platform
50. wavelength
51. world ocean
52. zooplankton

12.1 The Oceans

A science fiction author once noted that it is a funny thing that our planet is called Earth. Anyone looking at our planet from a distance would surely call it Ocean instead. Oceans cover about 71% of Earth's surface, giving Earth a distinct blue color when seen from space, as shown in the opening photograph of Chapter 1. Over half the people on Earth live within 200 kilometers of the ocean, but even if you live far inland, oceans affect your life every day, primarily through oceanic influence on the weather systems that move around the globe.

For most of human history, people have used the oceans for transportation and as a source for food. In the modern world, humans use the oceans for many

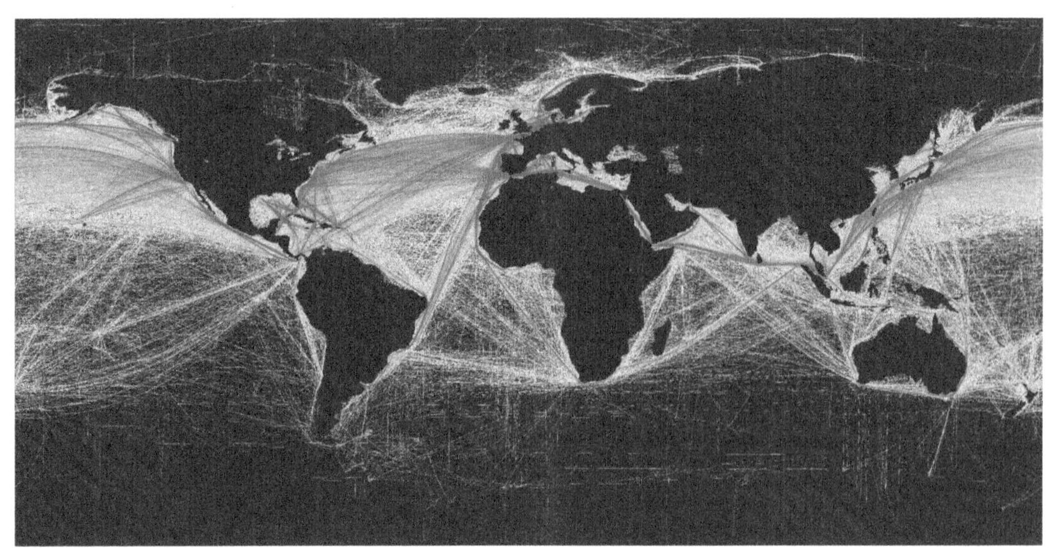

Figure 12.1. The red lines indicate shipping routes in the world's oceans. Greater shipping traffic generally means greater impact to the oceans.

things. The annual harvest of seafood is greater than 100 million metric tons.[1] Ten billion metric tons of cargo are shipped by sea every year, including shipments of everything from food and timber to cars and electronics. Figure 12.1 indicates the world's most frequently used shipping routes. About a third of worldwide production of oil and close to half of natural gas comes from rocks in the continental shelves. An increasing amount of our electricity comes from power plants that use the energy of offshore wind, waves, and tides. Another important use of the oceans is for recreation, such as swimming, snorkeling, and sailing.

We don't just use the oceans; at times humans abuse the oceans as well. Some vital fishing areas are overfished and seafood harvests in these areas are declining, a topic we addressed in Section 3.5.3. Pollution is a problem in many areas because the ocean is used as a place to dump garbage, raw sewage, and toxic wastes. Some species of fish contain high levels of toxins such as pesticides or mercury, making these fish unsuitable for human consumption.

Just as God calls humans to be wise stewards of land resources, such as water, soil, and energy resources, so he also calls us to be wise stewards of ocean resources. One part of this stewardship is to study the oceans. As noted in Chapter 1, the scientific study of the oceans is called *oceanography*. The men and women who conduct oceanographic research are called *oceanographers*.

12.1.1 History of Ocean Exploration

Having only primitive boats such as dugout canoes, early humans did not journey far out into the ocean; their voyages were mostly restricted to fishing trips with-

1 A metric ton is 1,000 kilograms, or 1 megagram. A metric ton is 2,200 pounds, slightly larger than an American ton, which is 2,000 pounds.

in sight of land. By 4000 BC, ancient civilizations in Egypt and Mesopotamia had developed sailing boats that could navigate across large seas out of sight of land, and by the first millennium BC, Greek, Phoenician, and Roman trading routes crisscrossed the Mediterranean. Later improvements in ship design enabled people to cross vast stretches of the oceans and by AD 1000, Viking ships had crossed the Atlantic Ocean to reach Greenland and even eastern Canada. Similar remarkable voyages were taken by people in the Pacific Ocean. Polynesian groups spread from island to island and developed methods for navigating to tiny atoll islands across vast stretches of ocean. Figure 12.2 shows a Polynesian ocean map, with sticks representing wave patterns and seashells representing islands.

Figure 12.2. Natives of some Pacific islands used stick charts such as this to plan voyages across vast stretches of ocean to reach tiny coral reef islands.

By the 1500s—during the European Age of Exploration—ships regularly traveled thousands of miles and the world was open for trade. These explorers had basic knowledge of the oceans but did not undertake any sort of scientific research. Scientific investigation of the oceans began in the 1700s, as traders and nations realized that they could profit by having a better understanding of the oceans. In the 1760s, Benjamin Franklin produced a map (shown in Figure 3.4) showing an ocean current called the Gulf Stream off the east coast of North America. By using knowledge of the Gulf Stream, mariners were able to cut several days off their travel time by either going with the current when going towards Great Britain or avoiding the current when travelling towards the American colonies. Ocean currents were further studied by U.S. Navy officer Matthew Maury in the mid-1800s—perhaps the first person to study the oceans as a profession. Maury built on Franklin's work to map ocean currents in other parts of the world and was the first person to describe worldwide surface wind patterns.

The first ship designated specifically for oceanographic research was the British ship *HMS Challenger*, pictured in Figure 12.3, which explored the Atlantic, Pacific, and Indian Oceans on a voyage that lasted from 1872 to 1876. The crew of the

Figure 12.3. HMS Challenger *sailed about 128,000 kilometers across Earth's oceans in the 1870s on one of the greatest scientific expeditions in history.*

Figure 12.4. The Ronald H. Brown is an oceanographic research vessel. NOAA is the U.S. government's National Oceanographic and Atmospheric Administration.

Challenger was able to grab samples of marine life and seafloor sediments from as deep as 8,000 meters, leading to the discovery of almost 5,000 new species of organisms. The scientists on board also measured ocean salinity (saltiness) and temperature and made observations about ocean currents and weather. The findings of the Challenger expedition were published in a 50-volume report that took over twenty years to write and publish, laying a foundation for the new science of oceanography. As is almost always the case, these scientists spent more time reading and writing than actually making observations and taking measurements.

Scientific investigation of the oceans accelerated in the 1900s with the invention of several new technologies. Before the invention of echo sounding (illustrated in Figures 2.26 and 2.27), the only way to determine depth was by using a rope or cable with a weight on the end. Satellites are now routinely used to take data representing surface water temperatures, wave heights and patterns, nutrient levels, and many other variables. Sophisticated oceanographic vessels, such as the research ship shown in Figure 12.4, cost tens or even hundreds of millions of dollars and are equipped with sophisticated scientific laboratories. Deep sea submersibles such as *Jason*, shown in Figure 12.5, are capable of descending to great depths to conduct scientific research.

Figure 12.5. Jason *is an unmanned, remotely operated vehicle that can descend to depths as great as 6,500 meters (21,400 feet). Jason can explore all parts of the ocean except the deep sea trenches.*

12.1.2 Oceans and Seas

Except for a few small inland seas, all the Earth's oceans are connected to one another. In a sense, there is just one *world ocean* surrounding scattered continents and islands. This one world ocean is divided into the more familiar named oceans—Pacific, Atlantic, Indian, and Arctic—and numerous smaller seas and gulfs, such as the Mediterranean Sea, the Caribbean Sea, and the Gulf of Mexico. The largest of the inland seas, which are not connected to the world ocean, is the Caspian Sea; other inland seas, such as the Dead Sea, are quite small. The world's oceans and major seas are charted in Figure 12.6 (placed on pages 340–341).

Figure 12.7. A Russian icebreaker cuts through sea ice near the North Pole. Icebreakers have especially strong hulls to protect the ships from damage caused by ice.

The Pacific Ocean is sometimes divided at the equator into the North Pacific Ocean and South Pacific Ocean. The same is true for the Atlantic Ocean. There is no real distinction between the northern and southern portions of these oceans and seawater freely mixes between them, just as water does throughout the world ocean.

To some oceanographers, the Arctic Ocean is just a northward extension of the Atlantic Ocean. The primary distinction of the Arctic Ocean is that large areas of its surface are covered by *sea ice*. Sea ice, with thickness up to 5 meters, is ice formed by freezing of seawater. The area covered by sea ice grows larger during the winter and shrinks during the summer. The area covered by summer sea ice in the Arctic Ocean is decreasing, and some scientists predict that the Arctic Ocean will be close to ice-free in the summer within a few decades. Figure 12.7 shows an icebreaker cutting through sea ice in the Arctic Ocean.

The ocean surrounding Antarctica also has sea ice, which extends well over 1,000 kilometers from the coastline of Antarctica during winter. Some maps show the ocean surrounding Antarctica as either the Antarctic Ocean or Southern Ocean; other maps show these waters as a southern extension of the Pacific, Atlantic, and Indian Oceans.

Learning Check 12.1

1. What do oceanographers mean when they speak of there being one world ocean?
2. What are some ways in which understanding the oceans is important for people in today's world?
3. What is the difference between an ocean and a sea?
4. Describe how the extent of sea ice changes from winter to summer in Arctic and Antarctic areas.

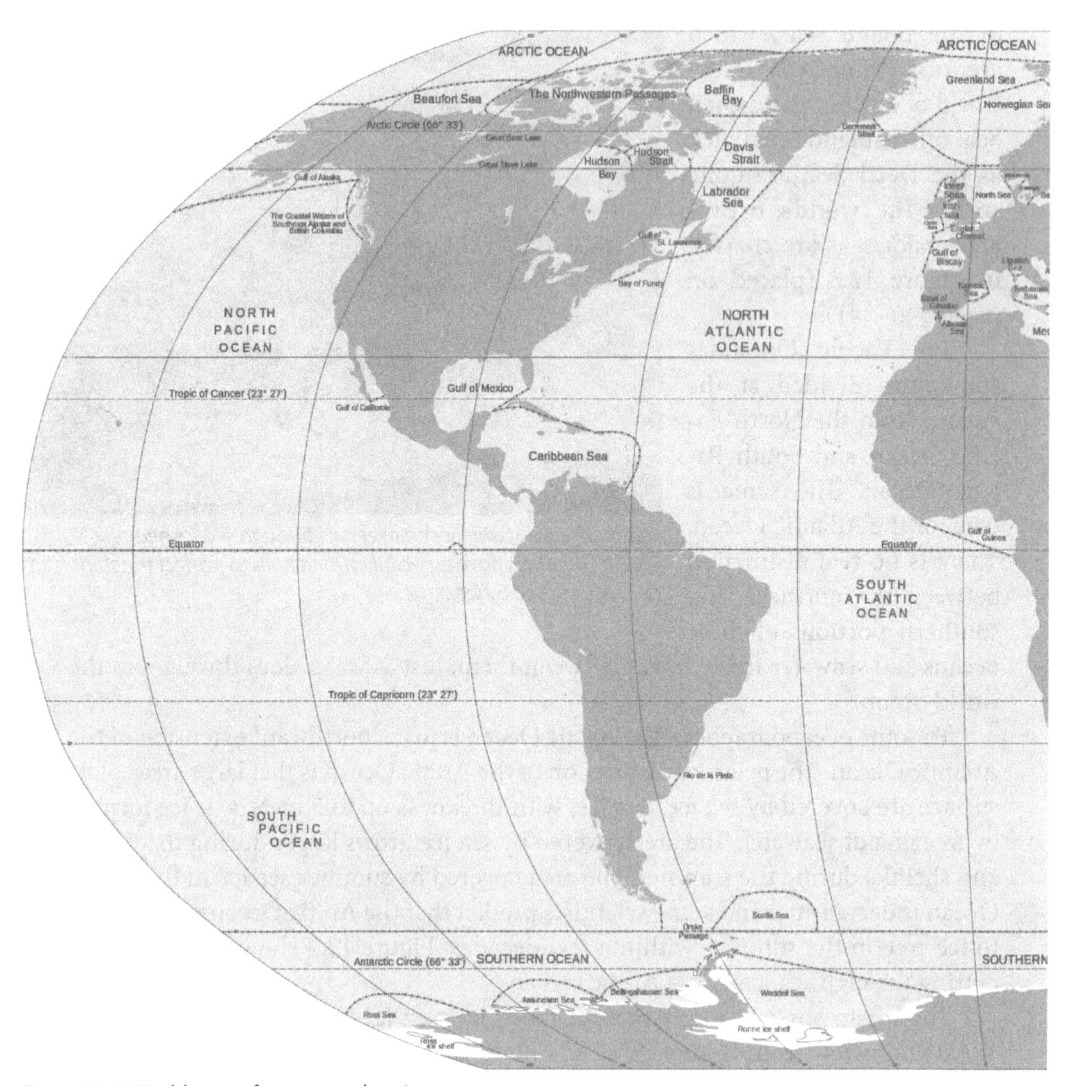

Figure 12.6. World map of oceans and major seas.

12.2 Seawater

As mentioned in Section 9.1, about 97% of Earth's water is in the oceans. Perhaps the most obvious characteristic of seawater is that it is salty. But seawater has other properties as well that are important to people who work in the ocean and that have a strong influence on the organisms that live in the oceans.

12.2.1 Salinity

Seawater is a solution composed of various salts dissolved in water, the most abundant of which is sodium chloride. In addition to sodium and chloride ions, seawater contains considerable amounts of magnesium, calcium, potassium, and sulfate ions, plus tiny amounts of almost every naturally occurring element in the periodic table.

The measure of the total amount of dissolved salts in seawater is called *salinity*. One kilogram of seawater typically contains about 35 grams of dissolved salts. Remembering that one kilogram is equal to 1,000 grams, you can calculate the percent of salt in seawater:

$$\text{salt percent} = \frac{35\text{ g salt}}{1000\text{ g seawater}} \cdot 100\% = 3.5\%$$

Instead of using percent, oceanographers usually express salinity in units of "parts per thousand," indicated with the symbol ‰. Calculating parts per thousand is just like calculating percent,[2] except you multiply by 1,000‰ instead of 100%:

2 The term percent actually means parts per 100.

$$\text{salinity} = \frac{35 \text{ g salt}}{1000 \text{ g seawater}} \cdot 1000‰ = 35‰$$

Obviously, a seawater sample can have a mass other than 1,000 g. For example, 3.5 g of salt in 100 g of seawater represents the same salinity as 35 g of salt in 1,000 g of seawater, or 0.7 g of salt in 20 g of seawater. The general formula for calculating salinity is:

$$\text{salinity} = \frac{\text{mass salt (g)}}{\text{mass seawater (g)}} \cdot 1000‰$$

Table 12.1 lists the abundances of the major constituents of seawater. Note that the concentrations of these major ions add up to almost 35‰. As already mentioned, seawater also contains tiny amounts of most other elements, such as iron, copper, and even gold. Because these elements are present in seawater in very small amounts, their concentrations are measured in units of parts per million (ppm) or parts per billion (ppb). The calculation of these values is similar to percent and parts per thousand. For example, 1,000 g of seawater contains about 0.0000000000013 g of gold, which is more easily expressed as 0.0013 ppb. Table 12.2 lists the concentrations of some minor components of seawater.

Component	Concentration (‰)
chloride (Cl^-)	18.9
sodium (Na^+)	10.6
sulfate (SO_4^{2-})	2.65
magnesium (Mg^{2+})	1.27
calcium (Ca^{2+})	0.40
potassium (K^+)	0.38
bicarbonate (HCO_3^-)	0.14

Table 12.1. Major dissolved ionic components of seawater.

Component	Concentration (ppm)	Concentration (ppb)
bromine	65	
strontium	8	
silicon	3	
nitrogen	0.280	280
zinc		110
iodine		60
phosphorus		30
iron		6
manganese		2
lead		0.04
mercury		0.03
gold		0.0013

Table 12.2. Minor elements in seawater.

The salinity of seawater varies from place to place, as indicated by the map in Figure 12.8. When seawater evaporates, only water molecules escape to the atmosphere; the dissolved salts stay behind. This leads to higher salinity in places with high rates of evaporation, especially around 20°–30° north and south of the equator where seawater salinity is typically around 37‰. Evaporation rates are high at the equator as well, but in most equatorial oceans, rainfall and evaporation are balanced, so seawater salinity remains at 35‰. In polar areas, melting ice from glaciers dilutes seawater and salinities are as low as 32 or 33‰. The lowest salinities in the oceans are near the mouths of large rivers, where seawater is diluted by the freshwater of the rivers. Some small seas have low salinities due to input from a large number of small streams. The salinity of the Baltic Sea in northern Europe is as low as 18‰. The highest salinities in the oceans are in smaller, subtropical seas that are mostly surrounded by land, such as the Red Sea between Africa and the Arabian Peninsula, where the salinity near the surface is as high as 40‰ in places. Even higher salinities are found in bodies of water in desert areas, such as the Dead Sea, where the salinity is over 300‰, nearly ten times as high as in the oceans.

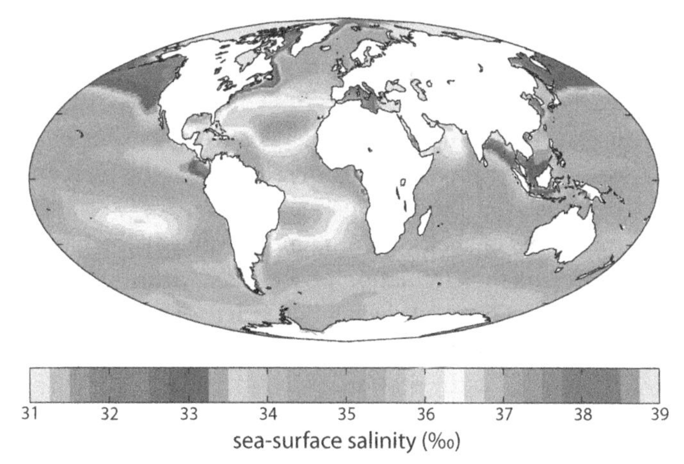

Figure 12.8. World map showing seawater salinity. Seawater with highest salinity tends to be in areas 20°–30° north and south of the equator, where the rate of evaporation is high and the amount of precipitation is low.

In addition dissolved salts, seawater contains dissolved gases such as oxygen, carbon dioxide, and nitrogen. A lot of the dissolved gases enters seawater from the atmosphere. Additional oxygen is produced by photosynthetic organisms, such as the seagrass shown in Figure 12.9. During photosynthesis, organisms take in carbon dioxide and produce oxygen. Marine organisms such as fish and invertebrates require oxygen in order to survive, just like land organisms. Some marine organisms, such as fish and clams, obtain their oxygen by using gills. Others absorb oxygen through their skin. Cold water holds more dissolved oxygen than warm water, so great numbers of organisms thrive in some polar waters.

Salts and other substances constantly flow into the oceans from rivers and other sources, but seawater doesn't become saltier over time. Just as there are processes that add salts to the ocean, there are processes

Figure 12.9. This seagrass in the Florida Keys is a plant, so it uses photosynthesis, but most of the photosynthesis in the oceans is performed by microscopic, single-celled organisms.

that remove salts. One of these is sea spray. When wind blows over the ocean surface near land, tiny droplets of water are carried inland, and each droplet contains a tiny amount of salt. Over time, this removes significant amounts of salt from the ocean, though much of this eventually washes back into the sea. A second process that removes ions from seawater is the formation of limestone. Limestone is composed of the hard parts of organisms. Sometimes these hard parts are large fragments of shells or coral; more typically the hard parts are microscopic. A third process is the circulation of seawater within oceanic crust, which is much like the movement of groundwater beneath the continents. Dissolved substances in the seawater react with hot rocks beneath the seafloor, changing the composition of both the rocks and the seawater.

12.2.2 Physical Properties of Seawater

We usually think of water as being transparent. We are able to see through a glass of water, under water from one side of a swimming pool to the other, or to the bottom of a clear stream or shallow lake. In small quantities, water is transparent because light travels through water for these short distances. However, there are limits to how far light is able to travel through water. Most sunlight does not penetrate deeper than about 100 meters beneath the ocean's surface, and below 200 meters depth the ocean is completely dark in all but the clearest water. Sunlight is composed of all the colors of the rainbow but not all colors travel the same distance through water. Blue light travels through water farther than other colors, so objects at depth appear blue. Seawater that contains suspended sediments (that is, it is muddy) is dark at much shallower depths.

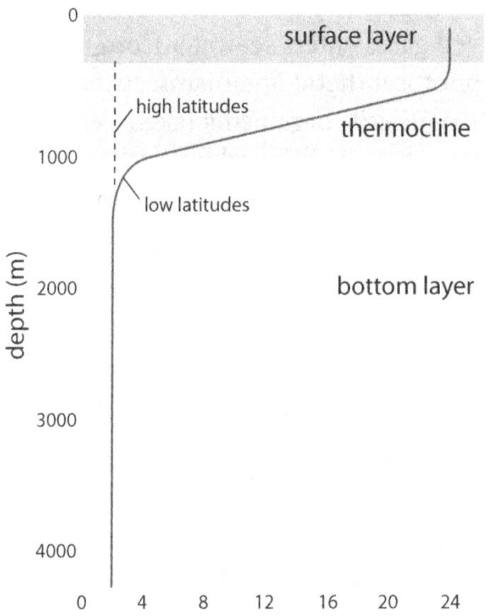

surface layer
high latitudes
thermocline
low latitudes
bottom layer
depth (m)
0
1000
2000
3000
4000
0 4 8 12 16 20 24
temperature (°C)

Figure 12.10. Seawater temperatures decrease with depth in tropical and mid-latitude oceans. In polar oceans, water temperature is cold from the surface to the bottom. The zone of greatest temperature change is called the thermocline.

When seawater freezes in Arctic and Antarctic areas, the resulting sea ice floating on the ocean surface is composed of fresh water. The remaining seawater is saltier and thus has a greater density than surrounding water, so it sinks to the bottom of the ocean. This very cold, saline water flows along the seafloor all the way to tropical regions. The result is that ocean bottom waters are cold worldwide, with a temperature of 2°C or colder, even at the equator.

In non-polar regions, the ocean has distinct layers, with cold, dense water at the bottom and sun-warmed water in the top 100 meters. The layer between deep, cold water and warmer surface is called the *thermocline*. Figure 12.10 shows how temperature changes with depth in most places in the ocean. In tropical areas, the ocean surface temperature is around 30°C. There is no thermocline in most polar waters; the surface water is as cold as –2°C. Figure 12.11 shows a map of the world with average sea surface temperatures.

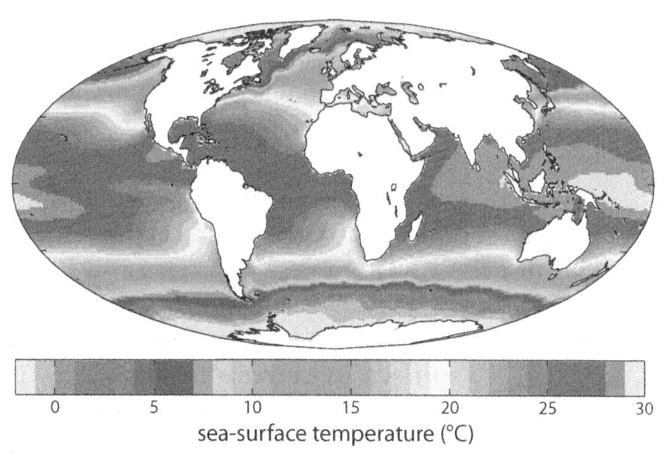

Figure 12.11. In general, average surface water temperatures are warm in the tropics and cold in polar areas, but there are important regional variations, such as off the west coast of South America.

Deeper in the ocean, the weight of all the overlying water is greater, which produces increased water pressure. You feel increasing water pressure on your eardrums if you swim to the bottom of a deep swimming pool. In a swimming pool, you only experience the weight of a few meters of overlying water; organisms on the abyssal plains and deep sea trenches experience the weight of several thousand meters of seawater. Most submarines operate at depths of just a few hundred meters and would be crushed by the pressures at greater depths. However, deep-sea organisms are adapted to these conditions and are not crushed by high pressures.

Learning Check 12.2

1. Describe the composition of seawater.
2. Why do oceanographers use units of ppm and ppb rather than ‰ for the seawater concentrations of elements like bromine and iron?
3. Why does seawater in the Arctic Ocean have lower salinity than in tropical oceans?
4. Explain why a thermocline is present in most of the world's oceans, but not present in polar areas.
5. Why is deep water colder than surface water in all but polar regions?
6. Why do most things appear bluish underwater?

12.3 Currents and Waves

Water in the oceans is in constant motion. This movement is caused by the sinking of dense polar waters, by gravity (in the case of tides), and by wind (in the case of waves and surface currents).

12.3.1 Ocean Surface Currents

An ocean *current* is a horizontal movement or stream of water in the ocean. There are different currents at different depths in the ocean, but in this section we will consider the currents on the surface. Surface currents move water in the uppermost 100 or so meters of the ocean.

It takes force to set an object in motion. In the case of surface currents, that force is supplied by winds that blow in consistent directions over large areas of the ocean's surface. In tropical regions, winds usually blow from east to west, creating surface currents in the ocean that also flow from east to west. Between 30° and 60° north and south latitude, winds usually blow from west to east, causing surface currents that also flow from west to east. If Earth were completely covered by oceans, the pattern of ocean currents would be very simple, with a band of water moving westward in the tropics, and bands of water moving to the east in mid-latitudes. But on the real Earth, surface currents are blocked by continents, causing the currents to deflect either northward or southward. The curved paths that currents follow is also due to a phenomenon called the *Coriolis effect*, which is caused by the rotation of Earth on its axis. The result of these factors is the striking pattern of ocean surface currents shown in Figure 12.12. In the North Atlantic Ocean and North Pacific Ocean, the Coriolis effect causes currents to move in clockwise loops called *gyres*. In the South Atlantic, South Pacific, and Indian oceans, the gyres move in a counter-clockwise direction. Each gyre is divided into four named currents, one flowing west, one north, one east, and one south.

Let's take a closer look at the surface currents in the North Atlantic Ocean. In Figure 12.12, we can start with the North Equatorial Current, flowing from east to west north of the equator. Being in the tropics, the North Equatorial Current absorbs a large amount of direct sunlight, so it is a warm-water current. As the current moves westward, it runs into South America, and is deflected northwestward toward North America. As the current moves north, its name changes to the Gulf Stream, which brings warm water up along the east coast of the United States. As the Gulf Stream moves northward, the water cools somewhat, but the

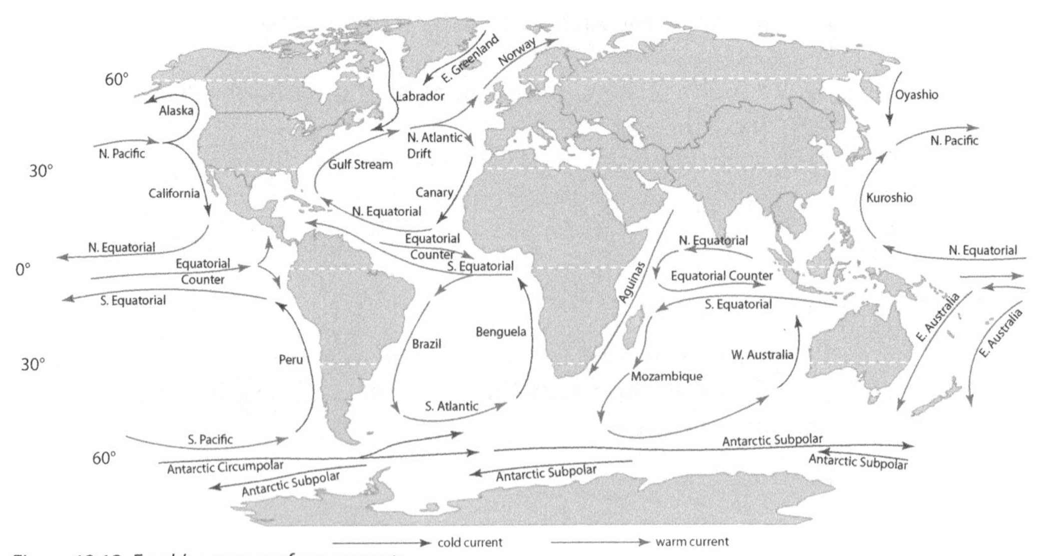

Figure 12.12. Earth's ocean surface currents.

Gulf Stream is still considered to be a warm-water current. The Gulf Stream is one of Earth's fastest-moving ocean surface currents, moving at speeds of up to 8 km/hr (5 mph) off the east coast of Florida. The volume of water moved by the Gulf Stream is about 300 times as great as the discharge of Earth's largest river, the Amazon. Figure 12.13 is a satellite image showing the warm waters of the Gulf Stream flowing northward in the Atlantic Ocean.

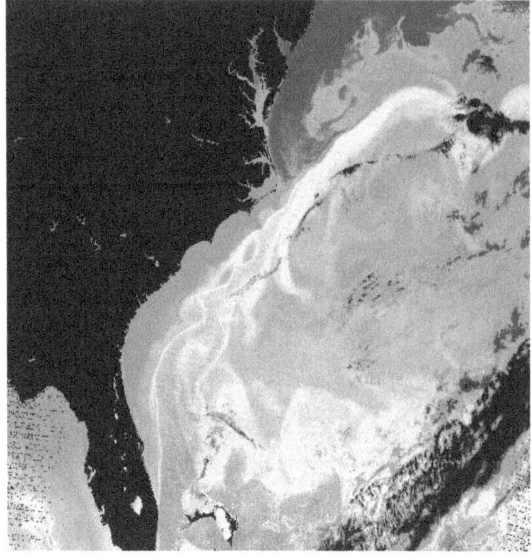

Figure 12.13. A sea surface temperature image of the Atlantic Ocean. Orange indicates warm water, blue indicates colder water, and yellow and green indicate water with intermediate temperatures. The Gulf Stream is the band of orange and yellow water off the east coast of the United States. The dark area in the southeast part of the image is a band of clouds.

As the Gulf Stream turns eastward, it becomes the North Atlantic Drift, carrying warm water across the Atlantic towards the British Isles and northwestern Europe. Even though the British Isles are as far north as central Canada, these warm currents keep the climate of the British Isles moderate. This area of moderate climate extends as far north as the central coast of Norway. The Canary Current, flowing down the west coast of Europe and North Africa, is a cold-water surface current. As the Canary Current flows southward, sunlight heats the water, and it eventually turns westward to complete the loop of the North Atlantic gyre.

The currents in the other gyres are similar. Warm currents flow away from the equator along the east coasts of Asia, Australia, South America, and Africa, and cold currents flow towards the equator along the west coasts of the Americas, Africa, and Australia. This difference between warm-water and cold-water currents explains why beachgoers comfortably swim in the Atlantic Ocean as far north as Long Island in New York, while the ocean off the coast of southern California is cold and swimmers there usually need wetsuits to stay in the ocean for any length of time.

12.3.2 Waves

In addition to causing large-scale surface currents, wind is responsible for most *waves* on the ocean surface. A wave is a rhythmic up-and-down or back-and-forth movement that carries energy through matter.[3] Water waves appear to move forward, but the actual path taken by water molecules or small objects in the water is circular, as illustrated in Figure 12.14. A cork floating on the water surface moves up and down and slightly forward then slightly back as waves pass by, but the cork

3 Or, in the case of light waves, through space.

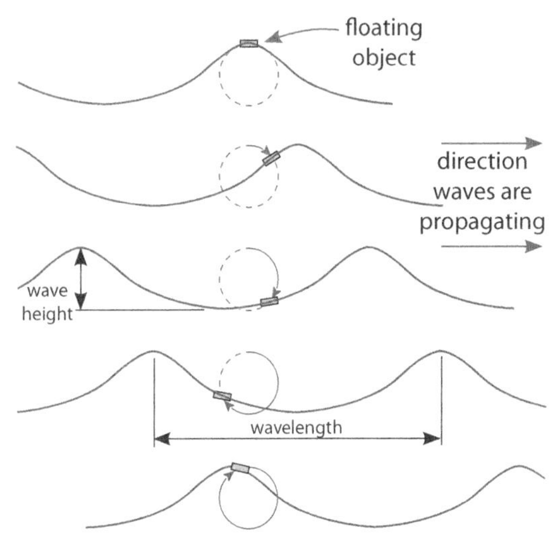

Figure 12.14. The circles indicate the movement of particles in the water as waves pass by.

itself not carried forward. The only thing that moves forward as a wave moves is energy.

Wind blowing over a smooth water surface causes small ripples to appear on that surface. Continued blowing of wind—especially wind blowing consistently in the same direction—causes those ripples to grow into larger and larger waves. The height of waves depends on three factors—wind speed, consistency of wind direction, and the distance or time over which the wind blows. If the speed or direction of the wind is inconsistent, smaller waves moving in different directions form instead.

A wave in deep water in the ocean or a lake has the features shown in Figure 12.14. The high points of the waves are called *crests*; low points are *troughs*. The horizontal distance from crest to crest or trough to trough is the wave's *wavelength*, and the vertical difference between the trough and crest is the *wave height*. The wave *period* is the amount of time it takes for one complete wave to pass by a location. For typical waves in the ocean or on a beach, the period is around 10 seconds, which means that a new wave passes a location or reaches the shoreline every 10 seconds.

The circular paths taken by water molecules become smaller with depth. Below a depth equal to about half a wavelength, the diameter of the circular paths becomes zero, and thus the wave energy is zero below that depth. Even in rough seas, the water at a depth greater than about half a wavelength is calm. When a wave moves from deep water to shallow water as waves approach a shoreline, the deeper orbits of the wave start to experience friction with the rocks and sediments on the seafloor. The waves slow down and grow taller and steeper. As waves move into even shallower water, the waves become yet taller and steeper and begin to tip over at the top, forming *breakers*, as shown in Figures 12.15 and 12.16. On a sandy or gravelly shoreline, each breaking wave surges up the beach face. The returning water forms a current beneath the breakers known as an *undertow*.

Figure 12.15. Breakers are important agents of erosion, sediment transportation, and deposition along shorelines.

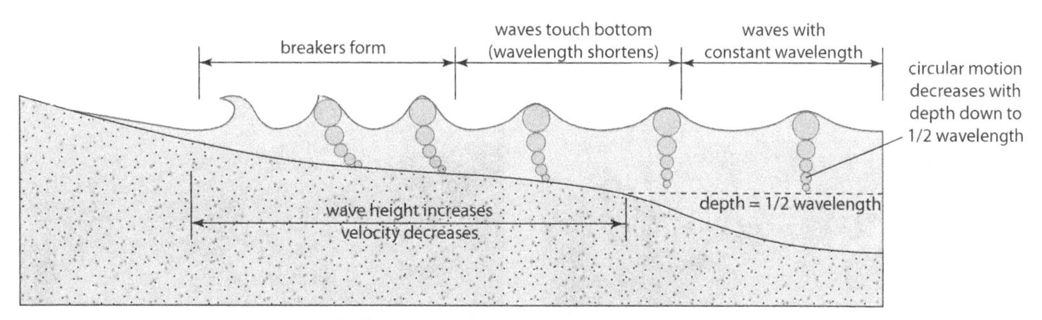

Figure 12.16. As waves move into shallow water, friction with the bottom causes wavelengths to become shorter and wave heights to become greater. Eventually the tops of the waves collapse, forming breakers.

Learning Check 12.3

1. Describe the motion followed by particles in seawater as a wave passes. How does this motion change with depth?
2. Draw a sketch of a wave with labels for the crest, trough, wavelength, and wave height.
3. What three factors determine the height of a wave?
4. Why do waves become breakers at shorelines?
5. What force causes surface currents to move?
6. Describe the currents that form the North Pacific gyre.
7. Explain why the water temperature at a beach in New Jersey is warmer than at a beach in southern California, even though New Jersey is farther north.

12.4 Tides

A *tide* is a periodic rise and fall of the level of the ocean surface at a particular place caused by the gravitational attractions between the Earth, moon, and sun. Figure 12.17 shows high tide and low tide in the Bay of Fundy in New Brunswick, Canada. At low tide, large areas of the sea floor are exposed to the atmosphere, while at high tide these areas are submerged beneath seawater. *Tidal range* is the difference in the height of the water surface between high tide and low tide. The Bay of Fundy has the greatest tidal range in the world, with a maximum 14.5 meter (47.5 feet) difference in elevation between low and high tides. Most shorelines around the world have tidal ranges of about 1–3 meters.

The ancient Greeks observed that there was some sort of connection between tides and the phases of the moon. In the 1600s, Isaac Newton gave a scientific explanation for tides using his theory of gravitation. Now we understand that the moon's gravity pulls on the solid Earth as well as on the oceans, but liquids deform much more easily than solids, so we notice ocean tides but not crustal tides. As Earth rotates on its axis, all areas of the oceans pass under the moon every 24 hours and 50 minutes. These gravitational attractions and the spinning of Earth cause a *tidal*

Figure 12.17. Boats at high tide and low tide in the Bay of Fundy, Canada.

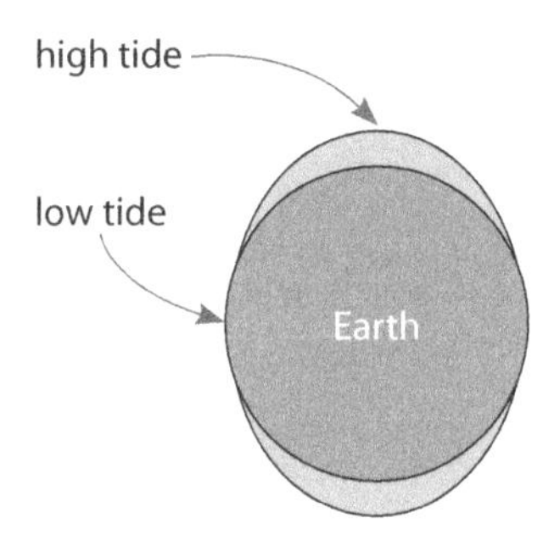

Figure 12.18. Simplified diagram showing tidal bulges on the sides of Earth closest and farthest from the moon.

bulge of seawater to exist on both the side of Earth facing the Moon and the side of Earth farthest from the moon, as illustrated in Figure 12.18. The gravitational attraction of the sun also influences tides, as we will see shortly.

If Earth were a smooth sphere completely covered by water, ocean tides would behave much as shown in Figure 12.18. However, Earth's landmasses interfere with the free movement of these tidal bulges as Earth rotates while the bulges move around the planet. The result is that some locations experience much greater tidal ranges than others.

The shapes of ocean basins and landmasses affect daily patterns of high and low tides as well. As shown in Figure 12.19, there are three general daily patterns of high and low tides. Many coastal areas on Earth, such as most of the Atlantic coast of the United States, experience two high tides and two low tides every day. Other coastlines, such as along the Gulf of Mexico, have only one high tide and one low tide per day. The twice-a-day pattern is re-

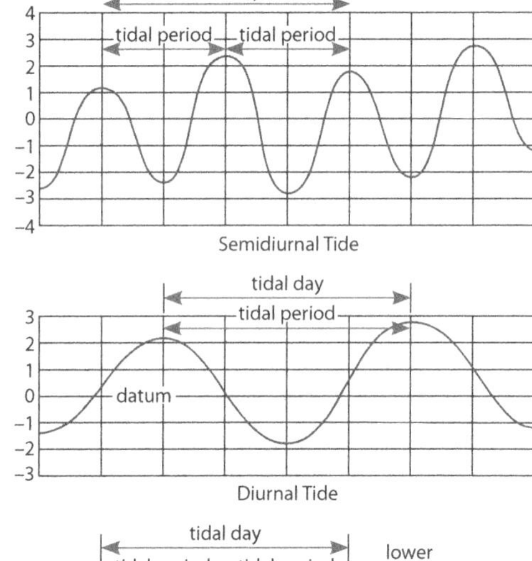

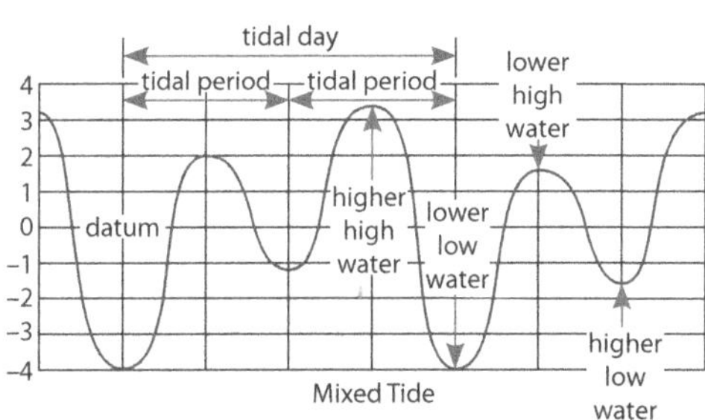

Figure 12.19. Semidiurnal tides have two roughly-equal high and low tides per day. Diurnal tides have just one high and one low tide per day. Mixed tides are a combination of the two other types.

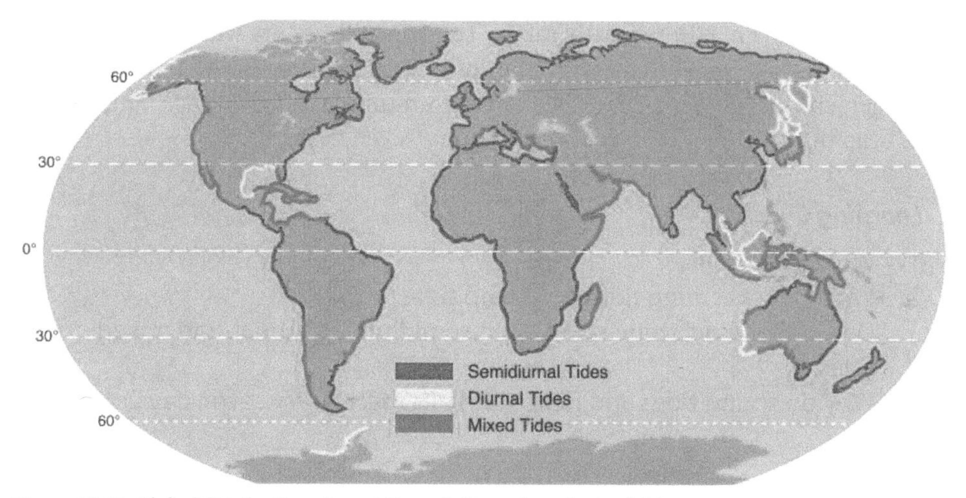

Figure 12.20. Global distribution of semidiurnal, diurnal, and mixed tides.

ferred to as *semidiurnal* and the once-a-day pattern as *diurnal.* Still other coastlines, such as most of the Pacific coastline of the United States, have a mixed tidal pattern with two high tides per day at significantly different heights. Figure 12.20 shows the global distribution of these three tidal patterns.

As mentioned earlier, the gravitational attraction of the sun also affects tides. Even though the sun is much larger than the moon, it is much farther away so the sun's influence on tides is less significant than the moon's. Depending on the phase of the moon, the combined influence of the moon's and sun's gravity can either build on each other, making tidal ranges greater, or oppose each other, making tidal ranges less, as illustrated in Figure 12.21. When the sun, moon, and Earth are all in a line—which occurs when there is either a full or new moon—the sun's and moon's gravitational attractions for Earth's oceans are also lined up, resulting in a *spring*

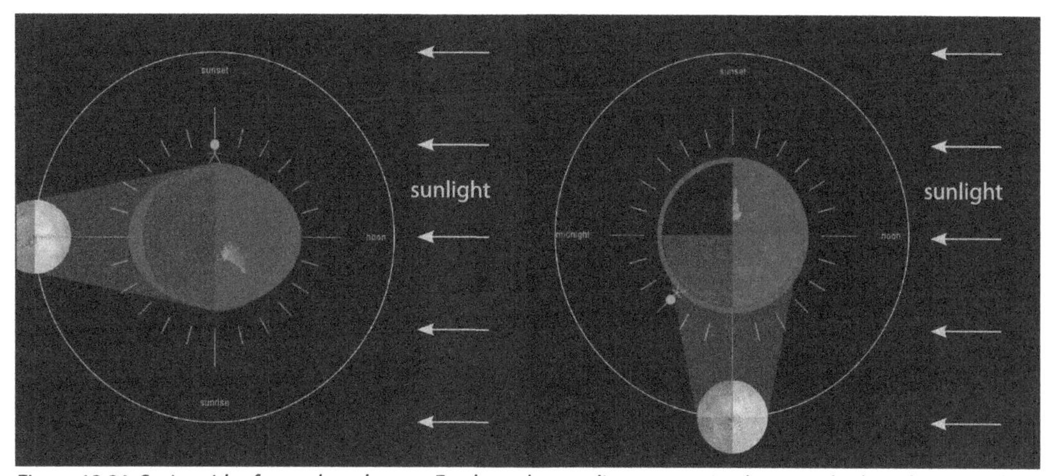

Figure 12.21. Spring tides form when the sun, Earth, and moon line up, causing lunar and solar tides to overlap and build on one another. Neap tides occur when the sun, Earth, and moon are arranged in an L-shape, which leads to smaller than normal tidal ranges.

tide, a tide with a maximum daily range. The opposite occurs when the moon is in its first or last quarter phase, when the sun, Earth, and moon are in an L-shaped arrangement. *Neap tides* are tides with a minimum daily tidal range. Spring tides and neap tides alternate with each other every seven days.

Learning Check 12.4

1. What causes tides?
2. What causes spring tides and neap tides?
3. Describe the differences between semidiurnal, diurnal, and mixed tides.
4. Why do spring tides and neap tides alternate every seven days?

12.5 Marine Life

Earth's oceans are filled with an astonishing variety of organisms, ranging in size from microscopic bacteria—the most abundant organisms in the oceans—to blue whales, possibly the largest animals ever to have lived on Earth.

12.5.1 Types of Marine Organisms

One way of grouping organisms in the oceans is according to how they move, which relates to where they live. The three major groups of marine organisms are plankton, nekton, and benthos.

Plankton are organisms that are either free-floating or weakly swimming. Most plankton are microscopic and they exist in vast quantities. A liter of near-surface seawater contains as many as 100 million planktonic bacteria, plus thousands of other planktonic organisms. There are two types of plankton. *Phytoplankton* are plant-like organisms that use photosynthesis. These microscopic marine phytoplankton process as much light and carbon dioxide through photosynthesis as all the plants on land! Figure 12.22 shows a type of phytoplankton known as *diatoms*—single-celled organisms enclosed in glassy cases. Animals, such as fish and invertebrates, do not make their own food like photosynthetic organisms do, so all marine animals are ultimately dependent on phytoplankton or photosynthetic plants for food.

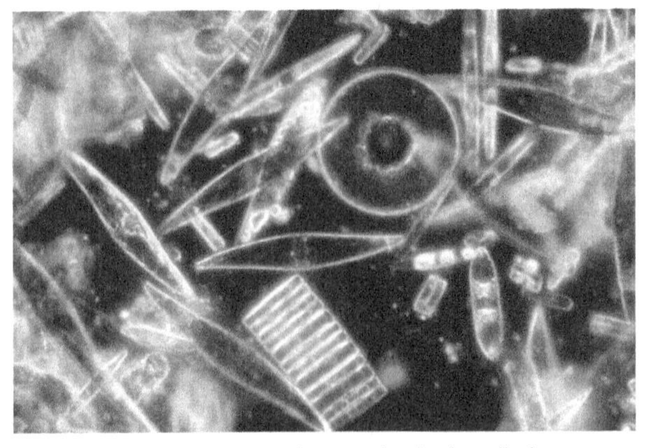

Figure 12.22. Diatoms are microscopic, single-celled phytoplankton enclosed in glassy shells. Diatoms are abundant in both marine and fresh water.

It is possible to measure the amount of photosynthesis that occurs in the ocean from space by measuring the amount of light

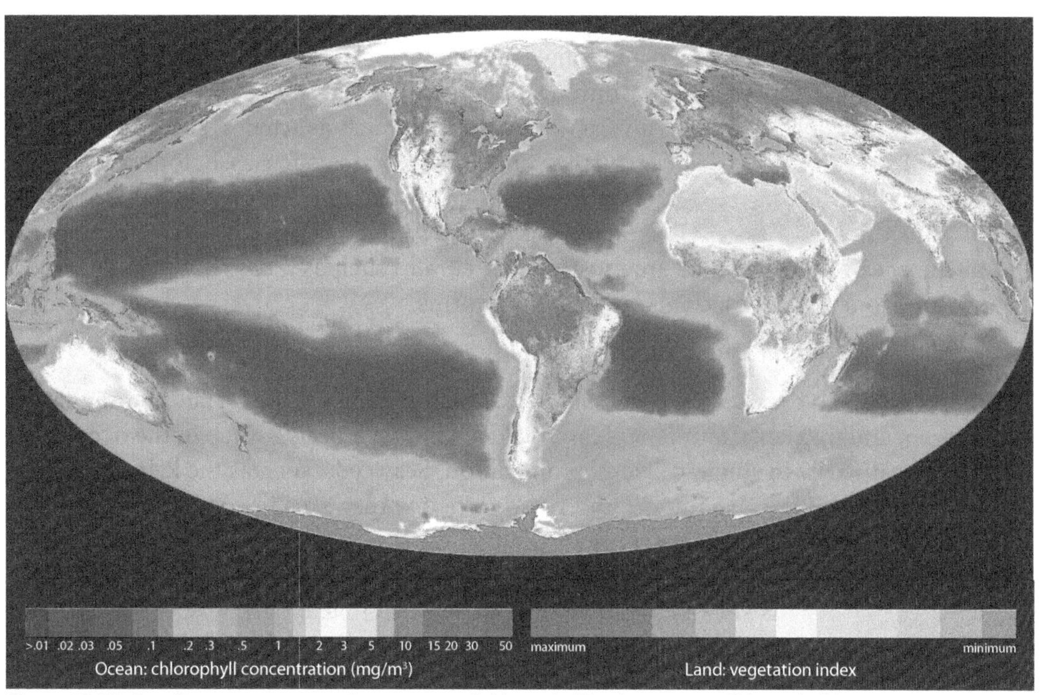

Figure 12.23. Satellites can measure the amount of chlorophyll in the oceans. Chlorophyll is a green pigment in plants and phytoplankton and is critically important for the process of photosynthesis.

absorbed by chlorophyll, a chemical compound that is essential for photosynthesis. Figure 12.23 is a world map of chlorophyll distribution based on satellite data. Areas that have more photosynthesis occurring have a greater amount of phytoplankton. Since all marine animals depend on photosynthetic organisms for food (as do land animals), a map showing where photosynthesis occurs is useful for showing where sea life in general is abundant and where it is sparse throughout the world ocean.

Animal-like plankton are referred to as *zooplankton*. Zooplanktonic organisms range in size from microscopic up to larger organisms such as jellyfish. Zooplankton feed on phytoplankton—the base for all food chains in the ocean—or on other zooplankton. Zooplankton are eaten in turn by larger organisms, such as fish and even some types of whales, such as blue whales. Figure 12.24 shows three types of zooplankton: *copepods*, the most abundant animal in the world; *krill*, shrimp-like

Figure 12.24. Three types of zooplankton: copepod, krill, and jellyfish. The copepod pictured is 1–2 mm long; the krill is several centimeters long; the bells of jellyfish range from 1 millimeter to 2 meters across.

organisms that are especially abundant in Antarctic waters; and jellyfish, creatures with a jelly-like body (or *bell*) and long, stinging tentacles.

Nekton are the actively swimming organisms of the ocean. Nekton include thousands of species of fish; a variety of mammals such as whales, sea lions, and dolphins; reptiles such as sea turtles; and invertebrates such as octopuses and squids. Nekton are most abundant in the uppermost 100 meters of the oceans.

The third group of marine organisms are the *benthos*, organisms that live on or in the ocean floor. Benthos are mostly invertebrates such as crabs, starfish, corals, sea anemones, sponges, clams, and marine worms.

12.5.2 Basic Requirements for Ocean Life

Most life in the ocean has the same basic requirements as life on land: sunlight, oxygen, and nutrients. Sunlight is essential for photosynthesis, which in the oceans occurs primarily in single-celled phytoplankton. Seaweeds, such as the kelp shown in Figure 12.25, are also important photosynthetic organisms in some marine environments. Photosynthesis is only possible in the sunlit uppermost 100 meters or so of the ocean, a layer known as the *photic zone*. The photic zone only contains about 2% of Earth's seawater; most seawater is in complete darkness. All animal life is dependent on photosynthetic organisms for food. Animals either directly eat photosynthetic organisms or eat something else that eats photosynthetic organisms. Because no photosynthesis occurs beneath the photic zone, deeper waters do not contain as many animals as shallow water. Deep-water organisms are mostly dependent on particles of food that slowly sink down from the photic zone.

Most living things, except for a few specialized types of single-celled organisms, require oxygen to live. The oxygen that fish and other marine animals (other than marine reptiles and mammals) breathe is dissolved in seawater. When a gas such as oxygen dissolves in water, it exists as individual molecules in between water molecules rather than as tiny bubbles. As organisms use oxygen, the oxygen is combined with carbon from food to make carbon dioxide. There are places in the oceans where there is little or no dissolved oxygen in the seawater, and few organisms can survive in these places. Most deep water in the Black and Caspian Seas, for instance, contains very little or no dissolved oxygen.

Figure 12.25. Kelp forests, such as this one off the coast of California, provide habitat and food for a great diversity of living things. Because sunlight is required for photosynthesis, kelp only grows in the photic zone.

A *nutrient* is a substance other than oxygen, carbon dioxide, or water that is required for organisms to live. On land, plants get their nutrients from soil; in the oceans, plants and phytoplankton get their nutrients directly from seawater.

Animals, whether on land or in the oceans, acquire their nutrients by eating other organisms. Essential nutrients in the oceans include nitrogen and phosphorus, and to a lesser degree metal ions such as iron, magnesium, and copper. If any one nutrient is in short supply, organisms are affected, and may even die in large numbers.

On the other hand, an overabundance of certain nutrients is harmful as well. Many rivers carry excess nitrogen and phosphorus from agricultural fertilizers into the oceans. These high levels of nitrogen and phosphorus are considered to be pollutants and they cause phytoplankton to multiply in vast numbers. The large numbers of phytoplankton consume most of the dissolved oxygen in the water, causing almost everything else to die and creating so-called "dead zones" in the oceans.

Organisms that live beneath the photic zone either rise up to shallower waters to feed at night or feed on nutrients that sink down from higher levels. Some cold, deep waters become nutrient-rich by accumulating these sinking nutrients. These cold, nutrient-rich waters are able to rise to the surface in some parts of the ocean through a process called *upwelling*. Upwelling occurs along some coastlines when steady winds move surface water further offshore. Deep water then rises to replace the surface water. Because of the availability of nutrients, areas of upwelling are some of the most productive commercial fisheries in the world.

12.5.3 Ocean Environments

The types of organisms found in any particular place in the oceans depend on a number of factors, such as water depth, distance to shorelines, water temperature, and amount of sunlight. The oceans can be divided into *benthic* environments, on the ocean floor, and *pelagic* environments, in the volume of the ocean above the ocean floor. There are a number of divisions within the benthic and pelagic environments, some of which are shown in Figure 12.26. Each of these zones has its own combination of environmental factors such as sunlight, water temperature, wave energy, and water depth, and is home to a different community of organisms.

Benthic environments are divided into zones based on depth. The shallowest benthic environment is the *intertidal zone*, located between high and low tide. The

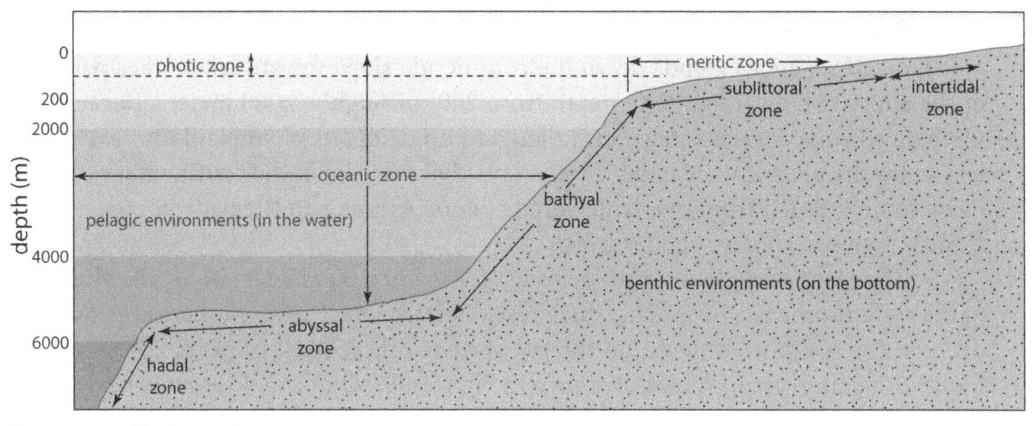

Figure 12.26. Marine environments.

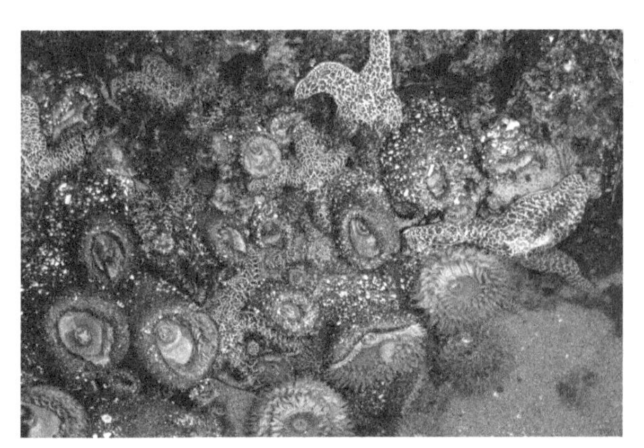

Figure 12.27. These starfish, sea anemones, and other creatures are regularly exposed to the atmosphere in a tide pool along a rocky coast.

intertidal zone is sometimes exposed to the atmosphere and direct sunlight, as shown in Figure 12.27. Due to breaking surf, the intertidal zone is a high-energy environment. Organisms that live in the intertidal zone are able to withstand pounding waves and being out of seawater for hours at a time. Organisms such as crabs and barnacles have hard shells that protect them from strong waves and predators as well as from drying out when exposed. A number of intertidal organisms, such as sea anemones and starfish, are either firmly attached to rocky surfaces or are able to hold on tightly. Many other intertidal species, such as clams, are able to burrow into sandy bottoms to protect themselves from waves, predators, and sunlight.

Beneath the intertidal zone is the *sublittoral* zone, extending out to the edge of the continental shelf. The seafloor of the continental shelf is rich with life. Some sublittoral animals move around on the seafloor in search of food. Others, such as corals, are attached to the seafloor and dependent on small organisms that float by for their food. Still others burrow into the sand. Because the sublittoral zone experiences relatively constant temperatures, currents, and sunlight, organisms do not have to deal with the constant changes that organisms in the intertidal zone do. Most of the sublittoral zone is in the photic zone, so there is an abundance of food from phytoplankton.

In some ways, we actually know more about the surface of Mars than we do about the deepest parts of Earth's oceans. Imagery taken from orbit around Mars has enabled scientists to map its surface in great detail, while Earth's oceans are hidden in darkness. Echo sounding and other techniques have allowed for mapping of the seafloor, but not in nearly the same degree of detail.

The *bathyal zone* extends down the continental slope towards the upper parts of the abyssal plains, ranging in depth from 200 to roughly 4,000 meters. Because the bathyal zone is completely dark, there are no plants or phytoplankton. Nutrients enter the bathyal zone either by currents that flow down the continental slope or by settling down from overlying sunlit waters. Animals in the bathyal zone include octopuses, sponges, and starfish.

Still deeper is the *abyssal zone* between 4,000 and 6,000 meters in depth. Much of the abyssal zone is empty of life due to lack of nutrients. However, there are oases deep on the seafloor, centered on the hot springs called black smokers. These are found along mid-ocean ridges, as shown in Figure 6.9. Microscopic organisms near the hot springs are able to use hydrogen sulfide in the hot water as an energy source to produce food, a process performed by photosynthesis in sunlit waters. The food produced by these organisms provides food in turn for a host of other organisms, such as clams, crabs, and the strange tube worms shown in Figure 12.28. The *hadal zone* occupies the deep sea trenches, the deepest part of the ocean. Little is known about the few organisms that live in the hadal zone.

Figure 12.28. Hydrothermal springs on the seafloor in the abyssal zone are home to giant tube worms, which are attached to the sea floor and grow to over 2 meters in length. Sea anemones and mussels are also visible in this photo.

The various benthic environments are all on the seafloor. All organisms which swim or float in the oceans above the seafloor live in the pelagic environment, which is divided into the neritic and oceanic zones. The *neritic zone* is the portion of the pelagic zone overlying the continental shelves. Most of the neritic zone is in the photic zone, allowing for an abundance of phytoplankton as well as larger photosynthetic organisms such as floating seaweeds. The neritic zone contains some of the greatest concentrations of marine organisms.

The great volume of the open ocean, beyond the continental shelves, is known as the *oceanic zone*. The upper part of the oceanic zone is also in the photic zone, and is home to a variety of fish, marine mammals, and invertebrates, but generally not in numbers as great as in the neritic zone. The deeper parts of the oceanic zone are dark and cold. Some whale species, such as sperm whales, spend most of their time near the surface but dive to great depths to feed on squid and other organisms that live deep in the oceanic zone. Most fish that live in constant darkness in the deepest parts of the oceanic zone are *bioluminescent*, which means they have body parts that glow in the dark. Examples are the various species of anglerfish, which dangle a lighted fin in front of their faces to lure smaller fish. (Go online and check these bizarre fish out!)

12.5.4 Ocean Food Resources

The oceans provide about 15% of the meat eaten by humans and have the potential to produce even more protein-rich food. But getting more food out of the oceans isn't as easy as build-

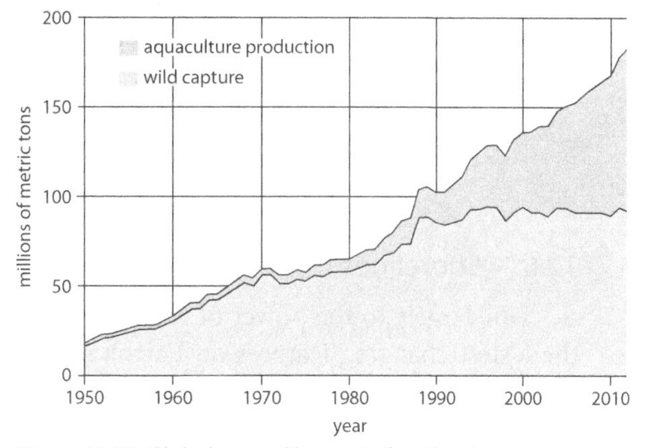

Figure 12.29. Global annual harvest of seafood.

Figure 12.30. These salmon pens in the North Atlantic Ocean represent just one of many types of aquaculture.

ing more fishing boats because many parts of the oceans are already over-fished. *Aquaculture* is the farming of the ocean, and is rapidly becoming an important source of food production. Aquaculture is already a significant source for salmon, tilapia, carp, and other fish, as well as for oysters, shrimp, and even seaweed for human consumption. Figure 12.29 shows that while traditional, wild capture of seafood has leveled off at around 100 million tons per year, the harvest from aquaculture has grown dramatically for the past few decades. Figure 12.30 shows large salmon pens with nets under water to keep the fish contained. Other aquaculture pens are constructed using ponds along coastlines or even onshore pools. One challenge for aquaculture is the massive amount of waste produced by corralling large numbers of fish or other marine animals in small spaces. This waste is a threat both to the fish in the pens, and to organisms in the surrounding ocean. Whether seafood is gathered from the wild, or grown in marine farms, it is important that the harvest is done in a way that is sustainable in the long term.

Learning Check 12.5

1. Distinguish between plankton, nekton, and benthos. Give at least two examples of each.
2. Why are phytoplankton essential to almost all other organisms in the oceans?
3. What things do marine organisms depend on in order to live?
4. Why do most marine organisms live in the photic zone?
5. Distinguish between benthic and pelagic environments.
6. Describe the five benthic zones.
7. Explain what aquaculture is and how it is being used to provide more food for humans.

12.6 Shorelines

Due largely to the power of waves, shorelines have landforms that are among the fastest changing features on Earth's surface. The constant pounding of waves accomplishes a great amount of erosion along shorelines and causes the continuous movement of sand along beaches. Erosion and deposition both contribute to the formation of the unique landforms found along Earth's coastlines.

12.6.1 Erosional Landforms along Coastlines

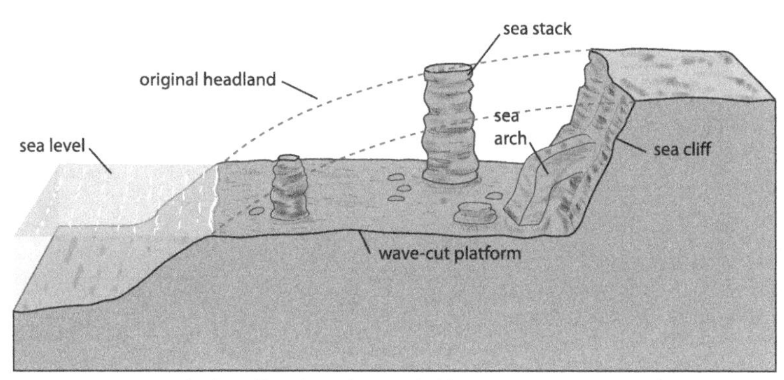

Figure 12.31. A rocky headland can be eroded by wave action to produce landforms such as sea stacks, sea arches, sea cliffs, and a wave-cut platform.

Rocky coastlines, such as the one shown in back Figure 12.15, experience constant beating from the force of breaking waves. These forces gradually break off pieces of weathered rock. As the surf moves the resulting gravel and sand back and forth, further erosion of bedrock occurs by abrasion. Because of their larger waves, storms are especially effective at eroding coastlines. The resulting erosional landforms are illustrated in Figures 12.31 through 12.33.

Figure 12.32. A sea stack and sea arch along the Atlantic coast of France. As waves continue to attack the shoreline, the stack and arch will eventually collapse and the cliff will retreat inland.

Waves are usually only a few meters in height, so most wave erosion occurs near sea level. Waves striking the base of a cliff typically wear a notch in the bottom of the cliff. As this notch grows deeper, landforms such as sea caves, *sea stacks*, and *sea arches* sometimes form. The ceaseless waves undercut these landforms, their structures collapse, and eventually they are eroded down to sea level. Over time, wave erosion cuts away more rocks and the shoreline retreats inland. Because waves do not erode below the level of low tide, this shoreline retreat results in the formation of a *wave-cut platform*.

Figure 12.33. A wave-cut platform and sea cliff in Wales at low tide.

12.6.2 Depositional Landforms along Coastlines

A *beach* is a sedimentary deposit of sand or larger fragments along an ocean or lake shore. When we think of beaches, we usually think of sand. But many beaches are composed of coarser material such as gravel or even boulders. Typically, sandy beaches are composed of quartz sand, with smaller amounts of other granitic minerals such as the feldspar minerals. But some beaches are composed of black or green sand derived from volcanic rocks. Still other beaches are composed largely of fragments of shells and coral. Some of the material for beaches comes from erosion

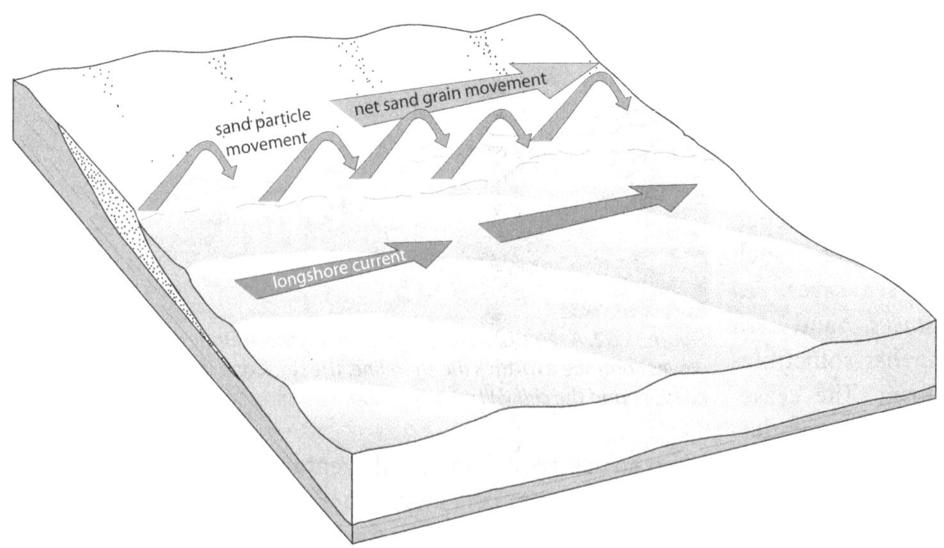

Figure 12.34. Waves usually do not strike a beach straight-on. As each wave goes up the beach face and then retreats, sand grains are moved in a zigzag pattern along the beach. The wave action also creates a current flowing parallel to the beach known as the longshore current.

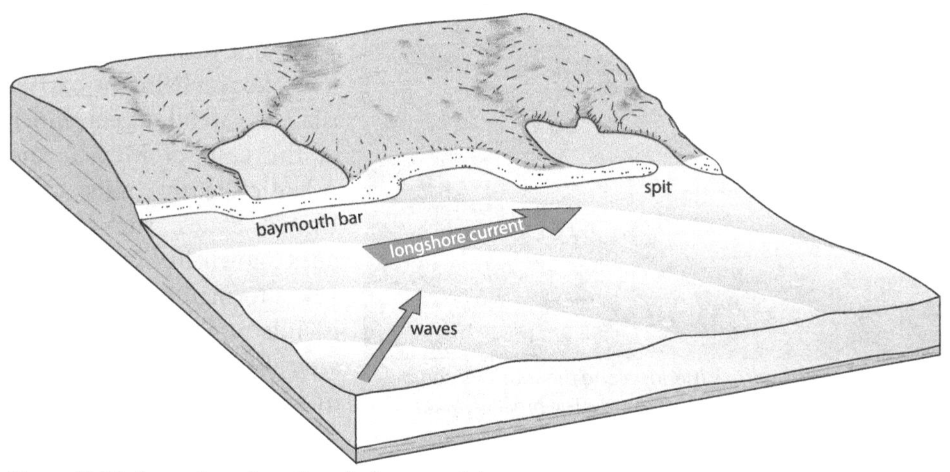

Figure 12.35. Formation of a spit and a bay mouth bar.

along rocky stretches of the coast. Other sediments for beaches are eroded off the continents and carried to the ocean by streams.

Waves not only move sand grains back and forth on a beach, but parallel to the shoreline as well. As illustrated in Figure 12.34, waves usually come into beaches at an angle, rather than hitting the beach straight on. With each wave, individual sediment grains are pushed a bit farther along down the beach, following a zigzag path in the general direction of wave movement. Waves striking the beach at an angle also give rise to a so-called a *longshore current* flowing parallel to the shore.

Longshore currents are responsible for creating characteristic landforms along shorelines, as illustrated in Figure 11.35. A *spit* is a narrow bar of sand that extends out from the shore into the ocean or into a bay. Figure 12.36 is a satellite image of a large spit in New Zealand. Spits grow longer over time or reach a length where they cannot extend any farther. If a spit extends all the way across a bay, it forms a *bay mouth bar*.

A *barrier island* is a long, narrow island that runs parallel to the shoreline. Longshore currents shape barrier islands and sometimes contribute to their formation. Barrier islands ex-

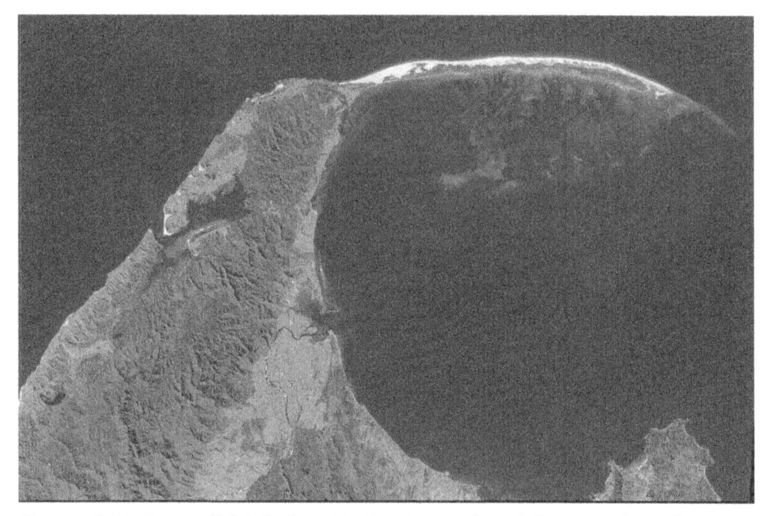

Figure 12.36. Farewell Spit in New Zealand extends 30 kilometers from the shoreline.

Figure 12.37. The long, narrow islands along the southern coast of Long Island, New York are barrier islands. The brackish lakes between the barrier islands and the larger island are lagoons.

ist up to several kilometers wide and tens of kilometers long. Long chains of barrier islands extend along much of the Gulf of Mexico and Atlantic Ocean coasts of the United States, such as the ones shown in Figure 12.37. Lakes formed behind barrier islands and bay mouth bars are called *lagoons*. The water in lagoons is usually *brackish*, which means the water is salty but not as saline as seawater.

Figure 12.38. A salt marsh, such as this one in Connecticut, is a type of tidal flat. Tidal flats are part of the intertidal zone.

A *tidal flat* is a muddy or marshy area covered and uncovered by the rise and fall of the tide. Tidal flats typically form in bays or behind spits, bay mouth bars, or barrier islands. When tidal flats are grassy, like the one shown in Figure 12.38, they are referred to as salt marshes. Tidal flats with saltwater-tolerant trees, as shown in Figure 12.39, are called *mangroves* or *mangrove forests*. (The same term is used for the trees found in mangroves.) All types of tidal flats are habitats for a diversity of organisms such as migratory birds, fish, and invertebrates. Tidal flats help protect inland areas from flooding during storms by providing a place to store excess water. When humans fill in tidal flats with earth for housing, industry, or farms, this natural flood protection is destroyed.

Figure 12.39. Unlike most trees, mangrove trees are able to tolerate living in the salty intertidal zone.

12.6.3 Humans and Coastlines

Depositional landforms such as beaches, bay mouth

Figure 12.40. Five hundred structures in this coastal community in Louisiana were destroyed by high water and waves during a hurricane. Many of the homes have since been rebuilt in the same place.

bars, and barrier islands are temporary and subject to continuous change. Despite this, millions of people throughout the world live very close to the ocean. Hazards facing communities along beaches and barrier islands include damage from waves and high water associated with storms, hurricanes or cyclones, and tsunamis. Figure 12.40 shows a beach community in Louisiana that was leveled by hurricane Rita in 2005.

Even aside from disasters such as hurricanes and tsunamis, coastal communities face threats from long-term changes to coastal landforms. Many beaches experience regular erosion of sand, causing the shoreline to move closer to buildings. In an attempt to control beach erosion in some areas, engineers build structures extending out into the ocean, such as the *groins* shown in Figure 12.41. A groin is a wall that traps sand, extending perpendicularly from the beach into the water. However, trapping sand along one part of a beach usually leads to greater erosion along a different part of the beach, as illustrated in Figure 12.42.

Figure 12.41. The groins along this beach in Spain trap sand along the beach.

Earth's climate has been warming over the past several decades, and this climate change, which we will address Chapter 15, is causing sea levels to rise. Higher temperatures around the globe cause sea level to rise for two reasons. Seawater, like other

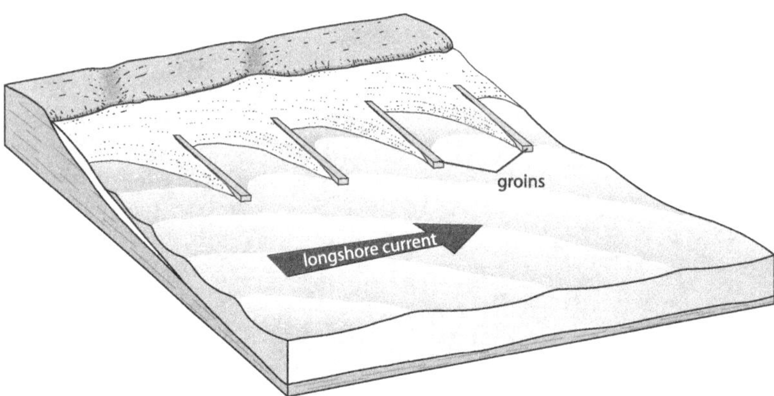

Figure 12.42. As the longshore current moves sand along a beach, sand is deposited on the up-current side of a groin and eroded on the down-current side.

substances, expands as the temperature increases, causing sea levels to rise. Higher temperatures also cause glaciers to melt faster than they form, contributing more water to the oceans and causing sea levels to rise. Higher sea levels lead to a greater risk of coastal flooding and erosion, especially during storms. For example, Miami Beach, shown in Figure 12.43, is only 1–2 m above sea level, and will be at increasing risk in the future. At present, sea level appears to be rising at a rate of 2.5 to 3.0 millimeters per year. This may not sound like much, but over a period of decades, this increase will threaten many low-lying coastal areas around the globe—and the people who live there.

Figure 12.43. Rising sea levels increase the risks of erosion, shoreline retreat, and flooding in many flat coastal areas. Miami Beach, Florida, is built on a barrier island, most of which has an elevation of only 1–2 meters above sea level.

Learning Check 12.6

1. Describe how sea stacks, sea arches, sea cliffs, and wave-cut platforms form.
2. What causes longshore currents and how do these currents affect sediments on beaches?
3. Explain how spits and bay mouth bars form.

Chapter 12 Exercises

Answer each of the questions below as completely as you can. Write your responses in complete sentences unless instructed otherwise.

1. Describe some technologies that have enabled oceanographers in the 21st century to study the oceans in much greater detail than was possible in the past. Include at least one technology not mentioned in this chapter.
2. Sketch map of the world with labels for its oceans and at least eight seas or major gulfs.
3. Why do you think that remotely-controlled submersibles are more practical for exploring the depths of the oceans than vessels designed to carry people?
4. What do you think has a greater density: seawater or freshwater? How could this hypothesis be tested?
5. An 80-g sample of seawater from the Red Sea contains 3.3 g of salt. What is the salinity of the sample in parts per thousand?
6. A seawater sample with a mass of 772 g is taken off the coast of Brazil. The water is evaporated from the sample and the remaining salts are weighed. If the mass of the remaining salts is 28.5 g, what is the salinity of the water in parts per thousand?
7. The concentration of gold in seawater is very small—a fraction of 1 ppb—but the amount of seawater on Earth is enormous. The result is that the value of all the gold in Earth's oceans is in the trillions of dollars. Why do you think no one bothers to extract gold from seawater?
8. The North Atlantic Drift is a slow, broad current, moving eastward at about 0.3 km/h. How many days would it take for a message in a bottle to travel from Halifax, Canada, to Cork, Ireland, a distance of about 4,100 km?
9. Speculate how the climate would change in England if the Gulf Stream and North Atlantic Drift stopped moving.
10. If wave erosion causes shorelines to move inland, why haven't the continents completely disappeared over time?
11. Irregular shorelines have headlands, which stick out into the ocean, and bays. The headlands experience greater rates of erosion than the bays. Predict what happens to these shorelines over time as the headlands are eroded.
12. If beaches and barrier islands are subject to occasional disasters, why do you think people build homes in such places? How might your answer be different for people building beachfront homes in a poor country instead of in the United States?
13. If sea level were to rise at a constant rate of 3.0 mm per year, how many years would it take for sea level to rise by 1.0 meter?

Chapter 13

The Atmosphere

The photograph above, taken by astronauts in the International Space Station, shows Earth's atmosphere as a thin blue layer surrounding Earth. We could ask many questions about this image, but a basic one is, "Why is the sky blue?"

Sunlight is composed of all the colors of the rainbow: red, orange, yellow, green, blue, violet, all the colors in between, plus ultraviolet and infrared radiation. When sunlight travels through Earth's atmosphere on a sunny day, most of the visible light travels directly to the surface. However, some of the light interacts with oxygen and nitrogen molecules in the atmosphere and is scattered in different directions. The short-wavelength colors—green, blue, and violet—are scattered much more than longer-wavelength colors—red, orange, and yellow. The result is that green, blue, and violet light come to our eyes from all directions of the sky, whereas red, orange, and yellow don't as much. Sunlight doesn't have much violet in it, so the scattered blue light makes the sky appear blue.

This scattering of short-wavelength colors also explains why sunsets are red. When the sun is on the horizon, sunlight travels through a greater thickness of atmosphere than when the sun is high in the sky. This greater thickness leads to even greater scattering of green, blue, and violet light, which means that the light directly from the sun that reaches your eye is more strongly composed of red, orange, and yellow.

In Chapters 13, 14, and 15, you will learn about Earth's atmosphere, the relatively thin layer of gases that make life possible on Earth. The atmosphere protects us from certain dangerous types of radiation from the sun and space, supplies oxygen for us to breathe, and providesmoisture for plants so we can eat.

Objectives

After studying this chapter and completing the exercises, you should be able to do each of the following tasks, using supporting terms and principles as necessary.

1. Describe the composition of Earth's atmosphere.
2. Compare and contrast the layers of Earth's atmosphere.
3. Describe atmospheric properties such as temperature, pressure, and humidity, and explain how these are measured.
4. Explain the processes of radiation, conduction, and convection, and how these processes are involved in the movement of heat energy in Earth's atmosphere.
5. Describe Earth's global patterns of wind.

Vocabulary Terms

You should be able to define or describe each of these terms in a complete sentence or paragraph.

1. absolute temperature
2. anemometer
3. atmospheric pressure
4. barometer
5. condensation
6. conduction
7. convection
8. dew point
9. doldrums
10. evaporation
11. exosphere
12. heat
13. heat index
14. horse latitudes
15. humidity
16. jet stream
17. mesosphere
18. meteorologist
19. meteorology
20. ozone
21. ozone layer
22. polar easterlies
23. polar front
24. precipitation
25. prevailing westerlies
26. radiation
27. radiosonde
28. relative humidity
29. sea breeze
30. stratosphere
31. temperature
32. thermosphere
33. trade winds
34. troposphere
35. wind-chill factor

13.1 The Composition and Structure of Earth's Atmosphere

The atmosphere is the layer of gas that surrounds Earth. The atmosphere is held in place by the gravitational attraction of Earth for the gas molecules in the atmosphere. The atmosphere is quite thin compared to the overall size of Earth—its thickness is comparable to the peel of an apple—but this thin layer makes it possible for life to thrive on our planet.

As mentioned in Chapter 1, *meteorology* is the scientific study of the atmosphere. A scientist who studies the atmosphere is a *meteorologist.*

13.1.1 Composition of the Atmosphere

Ancient Greek philosophers such as Plato and Aristotle taught that all things on our planet were composed of various mixtures of just four elements: earth, water, air, and fire. In this scheme, air was regarded as something that could not be broken down into simpler substances. Today we understand that Earth's atmosphere

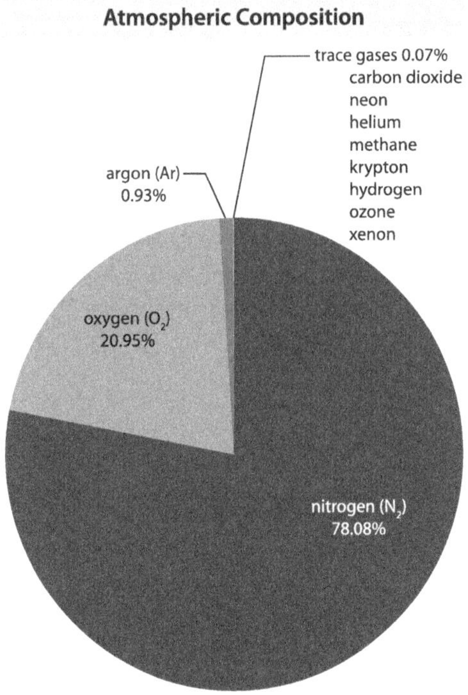

Atmospheric Composition

trace gases 0.07%
carbon dioxide
neon
helium
methane
krypton
hydrogen
ozone
xenon

argon (Ar) 0.93%

oxygen (O_2) 20.95%

nitrogen (N_2) 78.08%

Figure 13.1 The composition of Earth's atmosphere. (The amount of water vapor varies from 0 to 4 percent and is not shown.)

is composed of a number of different gases, each of which has its own properties.

Figure 13.1 shows the composition of Earth's atmosphere. The three most abundant gases in the atmosphere are nitrogen (N_2), making up about 78% of the gas molecules in the atmosphere, oxygen (O_2), making up about 21%, and argon (Ar), making up about 1%. In addition to these, there are trace amounts of a number of other gases, such as carbon dioxide (CO_2), which makes up about 0.04% of the atmosphere. Though these minor gases are present in only small amount, some of them, such as carbon dioxide and ozone (O_3), are still very important for life on Earth.

Another important atmospheric gas is water vapor. Water exists on Earth's surface in liquid, solid, and gaseous phases; water vapor is the gaseous form. Water vapor is invisible; clouds are made of microscopic droplets of liquid water or of ice crystals, not water vapor. The amount of water vapor in Earth's atmosphere, unlike the amounts of other gases, varies considerably from place to place and is continuously changing. Water vapor enters the atmosphere by evaporation from bodies of water and the land surface and is removed by condensation and precipitation. Very cold, dry air contains close to 0% water vapor, while hot, humid air contains levels as high as 4%.

The amount of carbon dioxide in Earth's atmosphere also varies over time, but not as quickly or dramatically as the amount of water vapor does. Carbon dioxide is a trace gas, with a concentration of about 0.04%.[1] Carbon dioxide is one of the so-called greenhouse gases, which means that it helps to control how much energy the atmosphere absorbs. This in turn controls the overall temperature of the atmosphere. We will explore the role of carbon dioxide in controlling the temperature of the atmosphere in Chapter 15.

Another important trace gas in Earth's atmosphere is *ozone*. Almost all oxygen in Earth's atmosphere is in the form of O_2: two oxygen atoms chemically bonded together in each molecule. The chemical formula of ozone is O_3: three oxygen atoms bonded together in each molecule in an L-shape. Most ozone in the atmosphere is in a region commonly called the *ozone layer*, typically between 20 and

1 This is usually expressed as 400 ppm (parts per million) rather than as 0.04%.

30 kilometers above Earth's surface. Even in the ozone layer, the
ozone is quite small—around 12 parts per million. But this tiny a
absorbs dangerous ultraviolet (UV) radiation from the sun, preven
radiation from reaching Earth's surface. Ultraviolet radiation is ha
to plants, animals, and even photosynthetic plankton in the ocea
sunburns and skin cancer in humans. Ozone in the ozone layer is ... by certain pollutants produced by humans, such as chlorofluorocarbons (CFCs) that were at one time widely used in refrigerants, hair sprays, and spray paints. In order to protect the ozone layer, the use of CFCs is now strongly limited by an international treaty.

A final important component of the atmosphere is microscopic particles of solids such as dust and salt. Dust becomes suspended in the atmosphere as wind blows over land areas; salt particles come from sea spray. These dust and salt particles are important for the formation of clouds because the tiny water droplets and ice crystals that make up clouds each form around one of these solid particles.

13.1.2 Layers of the Atmosphere

We describe the atmosphere in terms of layers, illustrated in Figure 13.2. These layers are primarily distinguished by how the temperature changes with increasing altitude, indicated by the red curve in the figure. (In contrast to the varying temperature, the gas composition of the lowermost 100 km of the atmosphere is fairly constant and exhibits the percentages shown in Figure 13.1.)

The lowermost layer of the atmosphere is the *troposphere.* This layer is characterized by a decreasing average temperature from ground level up to an average altitude of about 11 km, where the temperature stops decreasing with height. The troposphere contains most of the mass of Earth's atmosphere, as well as most of the water vapor. Most clouds are in the troposphere and the troposphere is where most weather phenomena develop. The top of the troposphere is called the *tropopause.* The altitude of the tropopause actually varies by latitude and from season to season.

The *stratosphere* begins at the tropopause and extends from about 11 km to close to 50 km above Earth's surface. The lower stratosphere con-

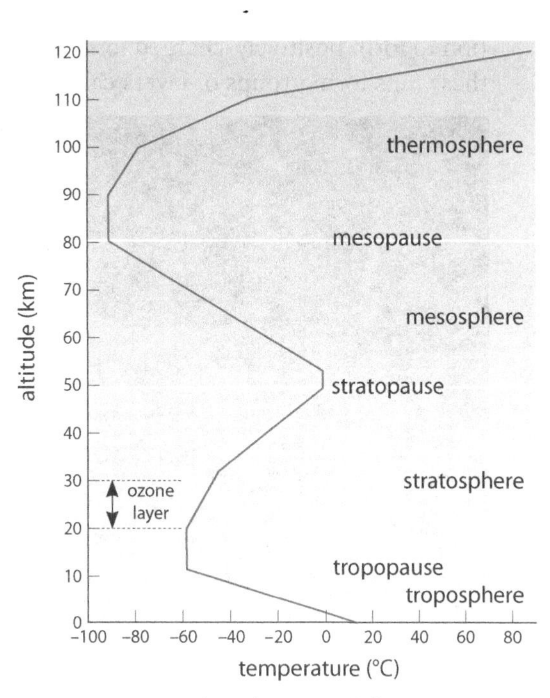

Figure 13.2. Atmospheric layers are defined by changes in temperature with altitude. Each increase and decrease of the temperature curve can be explained by how solar radiation interacts with gases in the atmosphere at different levels.

ins the ozone layer. The stratosphere is warmed by ozone absorbing ultraviolet solar radiation, resulting in an increasing temperature with increasing altitude. The top of the stratosphere is called the *stratopause*.

The *mesosphere* lies above the stratopause, between 50 and 80 km in altitude. The air at this level is extremely thin; 99.9% of all gas molecules in the atmosphere are below the mesosphere. The mesosphere does not contain higher levels of ozone, so the temperature again decreases with height. The top of the mesosphere is marked by the *mesopause*, the coldest part of Earth's atmosphere, with average temperatures around –90°C.

Above the mesopause is the *thermosphere*, the part of the upper atmosphere where temperatures increase with altitude. Gases in the thermosphere absorb energetic solar radiation such as X-rays and some ultraviolet radiation, causing temperatures to increase to as high as 2,000°C. Despite these high temperatures, the thermosphere does not "feel warm" because the air molecules and atoms are so far apart from each other that they do not effectively transmit heat to objects. Most satellites orbiting Earth, as well as manned spacecraft such as the International Space Station, pictured in Figure 13.3, are actually in the thermosphere. Over a period of many years, there is sufficient friction between the widely spread gas molecules of the thermosphere and a satellite to cause the satellite to slow down and drop out of orbit. The top of the thermosphere is around 500 km (300 mi) above Earth's surface.

Within the thermosphere, some gas molecules are broken apart by solar radiation to form positively-charged ions and free electrons. Within the thermosphere , these ions form groups of layers called the *ionosphere*. One reason the ionosphere

Figure 13.3. The International Space Station (ISS) orbits between 330 and 435 km (205 and 270 mi). The ISS experiences friction with the very thin thermosphere and must periodically be boosted to a higher orbit to prevent it from falling to Earth.

is important is because it reflects some wavelengths of radio waves, such as those from AM radio stations, transmitted from Earth back toward the surface, as illustrated in Figure 13.4. The ionosphere moves higher at night, allowing for even greater distances of radio signal transmission. The ionosphere is not a separate temperature layer, but a region of ion concentration within the thermosphere.

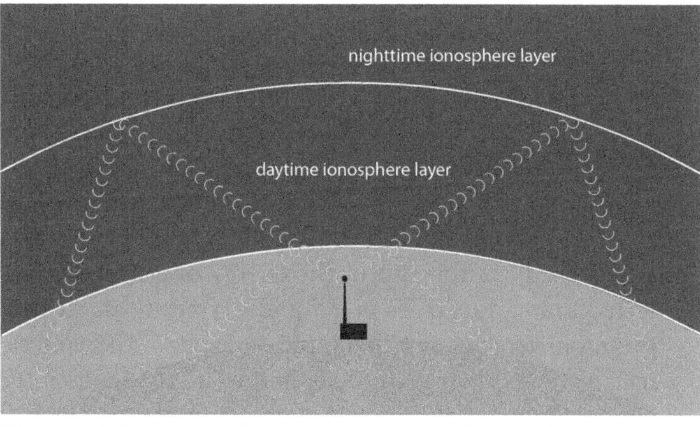

Figure 13.4. Some types of radio waves bounce off layers of the ionosphere, allowing them to travel great distances around the world.

The extreme upper layer of Earth's atmosphere is the *exosphere*, composed of light gases such as hydrogen and helium. The air molecules and atoms in the exosphere are so far apart that it is close to being an airless vacuum. There is no exact upper boundary to the exosphere; air molecules simply become farther and farther apart with greater height as the atmosphere merges with interplanetary space.

Learning Check 13.1

1. What are the three main gases found in Earth's atmosphere, in order of abundance?
2. What is ozone, where is most ozone found in Earth's atmosphere, and why is it so important?
3. What are the five main temperature-based layers of Earth's atmosphere, from bottom to top?
4. Explain the main factors causing the temperature variation with height in the stratosphere and thermosphere.

13.2 Properties of the Atmosphere

If you have watched a weather report on television, you have heard a meteorologist speak of high and low temperature, wind speed and direction, barometric pressure, and relative humidity. These are examples of measurable properties of the atmosphere. These properties are measured on the ground at weather stations and at high altitudes by weather instruments attached to weather balloons.

13.2.1 Temperature

If asked to give a definition of *temperature*, people might say that it is a measure of how hot or cold a substance is. This definition is in a sense correct, but a more precise definition is that temperature is a measure of the average speed of molecules or atoms in a substance. Fast-moving air molecules strike your skin with greater kinetic energy and you sense this as higher temperature. Conversely, slower-moving air molecules strike your skin with less kinetic energy and you sense this as air that does not feel as warm.

Temperature is commonly measured with a glass thermometer. Air molecules strike the glass walls of a thermometer with an amount of kinetic energy that depends on the temperature. The energy of fast-moving (warm) molecules is transferred to the molecules in a liquid—typically an alcohol with red dye—inside the thermometer, causing the liquid to expand; the volume of the red liquid is read on a scale as increasing temperature. If the temperature turns colder, air molecules striking the thermometer do not impart as much energy to the thermometer and the liquid contracts. There are other types of thermometers, such as electrical thermometers that measure electric currents flowing through certain materials. As the temperature increases, resistance to the flow of electricity decreases and this is measured and translated into temperature readings. Electrical thermometers are especially useful when there is no human observer to read the thermometer, such as at remote weather stations and in weather balloons.

Figure 13.5. The white box on the left is a weather instrument shelter housing thermometers, a barometer, and other instruments. The device to the right is a rain gauge.

In order to obtain accurate air temperature readings, it is critical that thermometers be placed in the shade. If a thermometer is placed in the sun, the thermometer produces a reading considerably higher than the actual air temperature. At official weather stations, instruments such as thermometers are kept in instrument shelters like the one shown in Figure 13.5.

Heat and temperature are related to each other, but they are not the same thing. *Heat* is energy being transferred from a warm object to a cold object. An object, such as a volume of air, can gain heat, which increases its temperature, or lose heat, which causes its temperature to decrease.

In the United States, temperatures in weather reports are typically given in degrees Fahrenheit (°F). However, in most of the world temperatures are reported in degrees Celsius (°C). In the mathematical equations meteorologists use to model the atmosphere for making forecasts, a temperature value must be expressed as *absolute temperature* on a temperature scale that has its zero point at *absolute zero*, the coldest temperature that is theoretically possible. The temperature of absolute zero is about -273°C (-460°F). At absolute zero, atoms and molecules are at rest, which means they also have zero kinetic energy. It is not possible for molecules to move more slowly than being at rest, so it is not possible for temperatures to be colder than absolute zero. The unit of measure used for absolute temperatures is the kelvin (K).[2] These three temperature scales are compared in Figure 13.6.

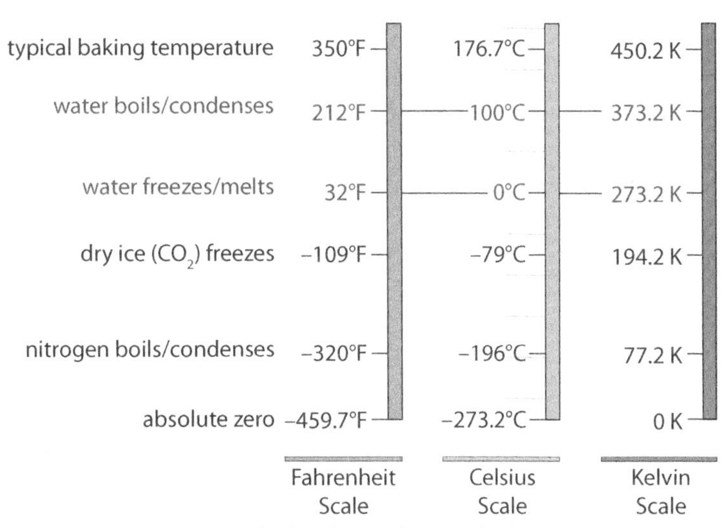

Figure 13.6. Comparison of Fahrenheit, Celsius, and Kelvin temperature scales.

In order to provide accurate weather forecasts, meteorologists must not only know the weather conditions on Earth's surface, but also at various levels within the troposphere and lower stratosphere. The most common way of obtaining these data is by using weather balloons like the one shown in Figure 13.7. Weather balloons support a box of instruments called a *radiosonde* that constantly sends out a radio signal with measurements of temperature, atmospheric pressure, and humidity.

Figure 13.7. A sailor prepares to release a weather balloon from a ship at sea. The instrument package in her hand sends weather data by radio.

2 The Kelvin temperature scale does not use the degree symbol, so absolute zero is -273°C, but 0 K.

13.2.2 Atmospheric Pressure

Molecules in the atmosphere are in constant motion. As they move, each air molecule in the troposphere collides with other air molecules billions of times every second. These molecules also hit other objects, such as Earth's surface and your skin. The combined force of all these collisions is what we call *atmospheric pressure.* Atmospheric pressure is the ratio of the average combined force of all the molecular collisions to the area the force pushes against. Thus, pressure is calculated as

$$\text{pressure} = \frac{\text{force}}{\text{area}}$$

One way to think of atmospheric pressure is that it is the weight of Earth's atmosphere pushing down on an area of Earth's surface as a result of Earth's gravitational attraction for that air. If we were to weigh all the air above a one-square-inch area at sea level, we would find the weight of the air to be about 14.7 pounds. This means that the average atmospheric pressure at sea level is 14.7 pounds per square inch.[3]

Higher in the atmosphere, the air is thinner—there are fewer molecules colliding with each other. As a result, atmospheric pressure always decreases with altitude. If you have ridden in a car in mountainous terrain, or if you have traveled in an airplane, you have experienced this lower pressure as a popping in your ears. As atmospheric pressure changes rapidly, the amount of pressure on the outside of your eardrums becomes greater or less than the air pressure on the inside of your eardrums. Your body equalizes that pressure when you swallow because of a narrow connection between your inner ear and the inside of your nose. Figure 13.8 shows how atmospheric pressure changes with altitude. Commercial aircraft have pressurized cabins so the crew and passengers are not exposed to the cold and low atmospheric pressures found near the top of the troposphere. Above 7,000 meters (23,000 feet), most mountain climbers must carry oxygen tanks because the air is significantly thinner.

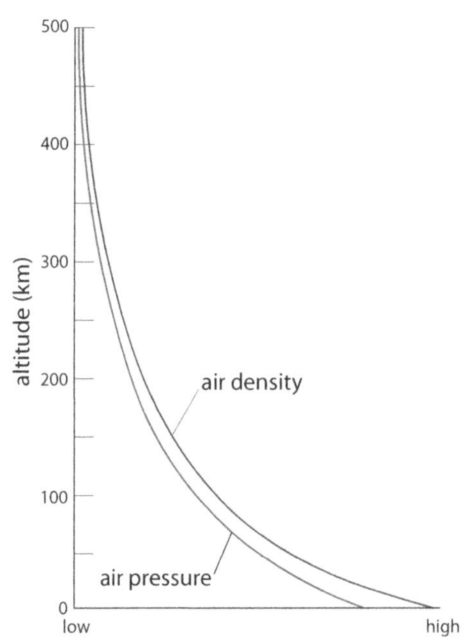

Figure 13.8. Both atmospheric pressure and air density decrease with altitude.

Atmospheric pressure also changes continuously at Earth's surface as various weather systems move around. These pressure

3 The metric equivalent to 14.7 pounds per square inch (14.7 psi) is 101,325 newtons per square meter (101,325 N/m^2), also known as 101,325 pascals (101,325 Pa).

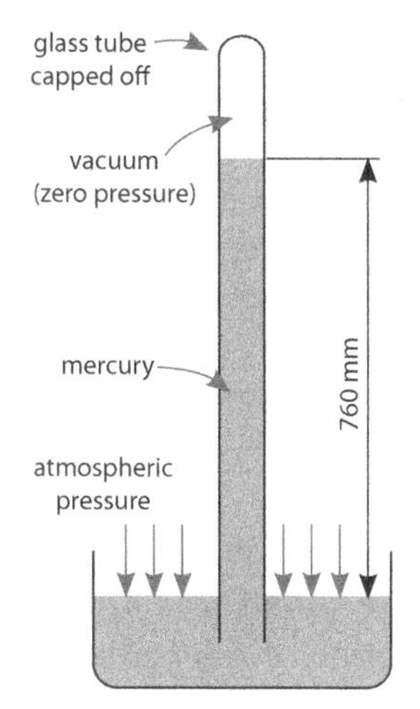

Figure 13.9. Atmospheric pressure is caused by numerous collisions of air molecules with the pool of mercury at the base of a mercury barometer. The higher the atmospheric pressure, the farther mercury climbs up the hollow glass tube.

changes do not happen as quickly as the pressure changes one experiences when changing elevation in an airplane, so our ears do not have to pop to compensate. Some weather systems entail higher atmospheric pressure and others entail lower pressure. High pressure systems usually come with clear skies and no precipitation, while low pressure systems are typically associated with clouds and precipitation.

Instruments used to measure atmospheric pressure are called *barometers*.[4] One type of barometer is a mercury barometer, illustrated in Figure 13.9. A mercury barometer consists of a vertical glass tube at least 84 cm (33 inches) long. The tube is closed at the top end, with an open, mercury-filled reservoir at the base. Mercury (Hg) is a metal that is a liquid at normal temperatures.

There is no air in the upper, closed end of the glass tube (a vacuum), so no force pushes down on the mercury column except gravity. The mercury is pushed up into the tube by atmospheric pressure pressing on the mercury in the bowl at the lower, open end of the tube. When atmospheric pressure increases—typical of fair weather—more pressure is exerted on the pool of mercury at the base and the column of mercury rises higher. When atmospheric pressure decreases—typical of stormy weather—the atmosphere exerts less pressure on the pool of mercury and the height of the mercury column decreases. On this barometer, the atmospheric pressure is expressed in terms of the height of the mercury column above the base, in units such as inches of mercury (in Hg) or millimeters of mercury (mm Hg). On a weather report, meteorologists typically state the current atmospheric pressure and the direction in which the pressure is changing, such as "29.87 inches of mercury and falling."

Figure 13.10. An aneroid barometer has a needle on a dial. The longer needle points to the current pressure, which on this barometer is given in inches of mercury and in millibars. The shorter needle is positioned manually every few hours, and is used as a reference to indicate if the pressure is rising, falling, or steady.

A more common type of barometer is the aneroid barometer,

4 Atmospheric pressure is sometimes called *barometric pressure.*

such as the one shown in Figure 13.10. An aneroid barometer uses a sealed metal box containing no air. As atmospheric pressure increases, it squeezes the box. The surface of the box is connected to a series of levers that control a needle on a dial, which moves on a scale. The pressure may still be expressed as inches of mercury, even though the aneroid barometer contains no mercury.

There are many other units used to express atmospheric pressure. Weather maps typically express pressure in millibar (mb), which are related to the unit newtons per square meter.[5]

13.2.3 Wind

Most of the air in Earth's atmosphere is in constant motion. Wind blows from areas of high pressure to areas of low pressure. These pressure differences are caused by the uneven heating of Earth's surface. A simple illustration of this is the formation of daytime *sea breezes* and evening land breezes along a coastline. Solar radiation heats up land more rapidly than water. On a sunny day, air above warm land heats up quickly, expands, and begins to rise. This creates a local area of low atmospheric pressure over the land. As air rises in one place, it must be replaced by air from another place. This creates a cool, gentle sea breeze that blows in from the ocean, as illustrated in Figure 13.11. The moist, rising air often produces a line of thunderstorms inland from the coastline. In the evening, the reverse occurs. The land surface cools more rapidly than seawater, and air begins to sink over the land, creating a warm land breeze that blows out over the water. The same process occurs on the coastlines of large lakes, forming lake breezes.

Usually we think of wind as blowing horizontally along Earth's surface. Air not only moves at the surface, but is in motion throughout the troposphere and in the lower stratosphere as well. Near Earth's surface, wind is slowed down by friction with the surface and with obstacles such as hills and trees, so wind usually blows faster higher in the atmosphere. This friction-caused boundary layer is up to one kilometer in thickness. Not only does wind blow horizontally, but air moves up and down in the atmosphere as well. Sometimes, such as during thunderstorms, these updrafts and downdrafts are more concentrated in small areas and blow quite strongly.

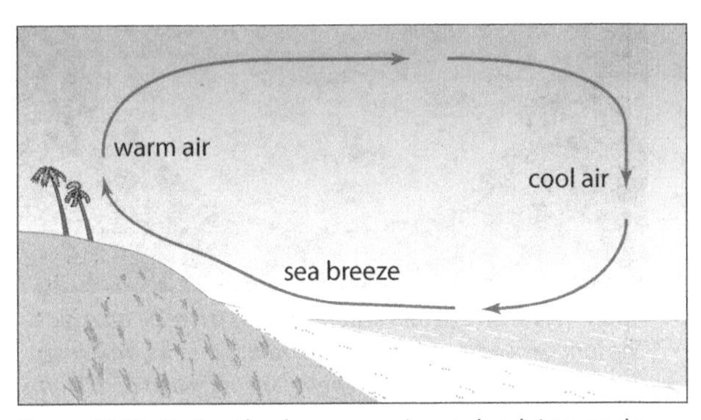

Figure 13.11. During the day, warm air over land rises, and a cool sea breeze blows in from the ocean.

Wind speed is measured with a device called an *anemometer*. A simple form of anemometer, shown in Figure 13.12, has cups on arms that rotate as the wind blows. In order to mini-

5 One millibar is one one-thousandth of a bar. One bar is equal to standard atmospheric pressure, equal to 14.7 psi = 101,325 N/m^2 = 101,325 Pa.

mize the effect of friction between wind and Earth's surface, anemometers are typically placed about 10 meters above the ground in areas with no trees, buildings, or other obstacles. Wind speed is measured in units of distance per time, such as kilometers per hour (km/h) and miles per hour (mph); or knots—nautical miles per hour.[6] Wind direction is expressed as the direction wind blows from, not the direction it blows toward. So a northeasterly wind blows from the northeast toward the southwest, and a south (or southerly) wind blows from the south toward the north.

Figure 13.12. An anemometer is used to measure wind speed. The faster the wind blows, the faster the cups rotate.

It is important to know the velocity of wind at higher levels of Earth's atmosphere as well as at ground level. A radiosonde cannot directly measure wind speed aloft because the weather balloon it is attached to is carried along by the wind. However, meteorologists track the position of the radiosonde by placing a GPS unit onboard or by analyzing its radio signals to determine its position. By tracking the position of the radiosonde, meteorologists are able to determine wind velocities at higher altitudes.

temperature (°F)

wind speed (mph)																		
calm	40	35	30	25	20	15	10	5	0	–5	–10	–15	–20	–25	–30	–35	–40	–45
5	36	31	25	19	13	7	1	–5	–11	–16	-22	-28	-34	-40	-46	-52	-57	-63
10	34	27	21	15	9	3	–4	–10	–16	-22	-28	-35	-41	-47	-53	-59	-66	-72
15	32	25	19	13	6	0	–7	–13	-19	-26	-32	-39	-45	-51	-58	-64	-71	-77
20	30	24	17	11	4	–2	–9	–15	-22	-29	-35	-42	-48	-55	-61	-68	-74	-81
25	29	23	16	9	3	–4	–11	–17	-24	-31	-37	-44	-51	-58	-64	-71	-78	-84
30	28	22	15	8	1	–5	–12	-19	-26	-33	-39	-46	-53	-60	-67	-73	-80	-87
35	28	21	14	7	0	–7	–14	-21	-27	-34	-41	-48	-55	-62	-69	-76	-82	-89
40	27	20	13	6	–1	–8	–15	-22	-29	-36	-43	-50	-57	-64	-71	-78	-84	-91
45	26	19	12	5	–2	–9	–16	-23	-30	-37	-44	-51	-58	-65	-72	-79	-86	-93
50	26	19	12	4	–3	–10	–17	-24	-31	-38	-45	-52	-60	-67	-74	-81	-88	-95
55	25	18	11	4	–3	–11	-18	-25	-32	-39	-46	-54	-61	-68	-75	-82	-89	-97
60	25	17	10	3	–4	–11	-19	-26	-33	-40	-48	-55	-62	-69	-76	-84	-91	-98

Frostbite occurs in: 30 min | 10 min | 5 min

Figure 13.13. Wind-chill equivalent temperatures. A 20 mph wind combined with an air temperature of 10°F makes it feel like –9°F, placing people at increased risk of frostbite.

6 A nautical mile is about 6,076 feet, 15% longer than the standard statute mile (5,280 feet).

On a cold day, the air feels even colder if the wind is blowing. Moving air is more efficient at removing heat from objects, such as your skin, than still air is. The *wind-chill factor* is a value calculated to indicate the cooling effect of wind. When wind is blowing, air does not actually become colder but it feels colder because of the loss of body heat. The chart in Figure 13.13 shows wind-chill factor values.

13.2.4 Humidity

Humidity is a measure of the amount of water vapor in the air relative to the maximum amount possible. There is a limit to how much water vapor air can hold. Warm air is able to hold up to 4 percent water vapor; cold air holds less. When air holds as much water vapor as it is capable of at a given temperature, it is said to be saturated. As air cools overnight, the air becomes more humid because the cooler air cannot hold as much moisture. As the air continues to cool, it may eventually reach the temperature at which the air is saturated with water vapor. This temperature is called the *dew point*. As the air cools below the dew point, water vapor condenses on cool surfaces in the form of dew, as shown in the lovely image in Figure 13.14. If the surface temperature cools to below the freezing point of water, 0°C, water vapor deposits on the surface as frost rather than as dew. Condensation may also take the form of microscopic water droplets in fog or clouds.

13.14. Dew forms when surfaces cool down to temperatures below the dew point by radiating heat at night.

Humidity is usually expressed in terms of *relative humidity*, defined as the ratio of the amount of water vapor in the air to the maximum amount of water vapor the air can hold at that temperature. Relative humidity is expressed as a percentage. For example, one cubic meter of air at 30°C is capable of holding a maximum of about 30 grams of water vapor. If this cubic meter of air only contains 15 grams of water vapor, the relative humidity is 50% because the air has only half as much water vapor as it is able to hold. If air is saturated, the relative humidity is 100%. It is important to remember that even when relative humidity is near 100%, water vapor still makes up at most 4% of the molecules in Earth's atmosphere. In deserts, daytime relative humidity in the summer is typically below ten percent. The driest place on Earth's surface is actually the area around the South Pole in Antarctica, where relative humidity measurements are typically near zero.

The *heat index* is similar in concept to the wind-chill factor, except it combines temperature and relative humidity to determine the temperature that humid air actually feels like. Heat index values are shown in Figure 13.15. The primary way

temperature (°F)

relative humidity (%)	80	82	84	86	88	90	92	94	96	98	100	102	104	106	108	110
40	80	81	83	85	88	91	94	97	101	105	109	114	119	124	130	136
45	80	82	84	87	89	93	96	100	104	109	114	119	124	130	137	
50	81	83	85	88	91	95	99	103	108	113	118	124	131	137		
55	81	84	86	89	93	97	101	106	112	117	124	130	137			
60	82	84	88	91	95	100	105	110	116	123	129	137				
65	82	85	89	93	98	103	108	114	121	128	136					
70	83	86	90	95	100	105	112	119	126	134						
75	84	88	92	97	103	109	116	124	132							
80	84	89	94	100	106	113	121	129								
85	85	90	96	102	110	117	126	135								
90	86	91	98	105	113	122	131									
95	86	93	100	108	117	127										
100	87	95	103	112	121	132										

Awareness of heat-related disorders with prolonged exposure or strenuous activity:

caution — extreme caution — danger — extreme danger

Figure 13.15. Heat index temperatures. A relative humidity of 70% combined with an air temperature of 90°F makes it feel like 105°F, placing people at increased risk of sunstroke or heat exhaustion.

the human body cools on a warm day is by sweating. It takes heat to make sweat evaporate and most of that heat comes from a person's skin, so when sweat evaporates heat is removed from the body. Sweating works most efficiently when the relative humidity is low. If the temperature is 100°F but the humidity is low, a person may actually feel cold when he climbs out of a swimming pool because the water evaporates quickly from his skin, removing a tremendous amount of heat from the body. On the other hand, on a hot day with high relative humidity evaporation occurs slowly because the air is already near saturation and cannot hold much more water vapor. The result is that sweat does not efficiently cool one's body on a hot, humid day.

13.2.5 Precipitation

Precipitation is any form of water—liquid or solid—that falls from the atmosphere and reaches Earth's surface. Rainfall is measured in millimeters or inches using a rain gauge, such as the one shown in Figure 13.16. Simple rain gauges have a funnel collector attached to a long measur-

13.16. A simple rain gauge can be built with an empty can and a ruler. This sophisticated rain gauge has metal slats to protect the opening from wind.

ing tube. As it rains, the tube fills with water and rainfall totals are read from a scale on the side of the tube. More sophisticated rain gauges include tipping-bucket rain gauges that have a small bucket to dump the water after an accumulation of 0.01 in of rain, and weighing-pan rain gauges that weigh rainfall and convert the weight to millimeters or inches.

Snowfall is more difficult to measure because wind blows the snow around, leading to various depths in a small area, and because snow quickly recrystallizes or compacts as it falls. Snowfall is typically measured in at least three areas at a weather station and the depth measurements are averaged. After measurement of snow depth, the snow is melted to determine the equivalent amount of rainfall. A foot of snow might melt down to anywhere from a half-inch to two inches of rain-water equivalent, depending on whether it is "wet" snow or "dry" snow.

Learning Check 13.2

1. Distinguish between heat and temperature.
2. What is absolute zero and what is its value in three different temperature units?
3. What causes atmospheric pressure?
4. How does atmospheric pressure change with altitude in the atmosphere?
5. Explain how sea breezes and land breezes form.
6. What is the difference between relative and absolute humidity?
7. Why does it feel warmer on a hot, humid day than on a hot, dry day?

13.3 Energy and Water in the Atmosphere

All weather is driven by the interactions of radiation from the sun with the four Earth systems: the atmosphere, the hydrosphere, the biosphere, and solid Earth. As weather systems form and move, energy is transferred from one place to another by three means: radiation, conduction, and convection.

13.3.1 Solar radiation and Transfers of Energy

Radiation is the transfer of energy by electromagnetic waves. Recall from Chapter 2 that the electromagnetic spectrum includes radio waves, microwaves, infrared light, visible light, ultraviolet light, X-rays, and gamma rays (illustrated in Figure 2.21). Unlike sound waves, water waves, or seismic waves, electromagnetic waves do not need a physical medium in which to propagate, so they are able to travel through the vacuum of space.

When solar radiation hits Earth's atmosphere, some wavelengths of radiation are absorbed by the atmosphere and do not penetrate all the way to Earth's surface. We have already seen that Earth's thermosphere absorbs X-rays coming from the sun and that the ozone layer absorbs most of the ultraviolet radiation. Clouds reflect some visible and infrared radiation back into space. Incoming solar radiation that is not absorbed or reflected by the atmosphere is either absorbed by Earth's surface, resulting in warming of the surface, or reflected by Earth's surface, in which case the radiation returns to space. Earth's warm surface then re-emits infrared radiation.

Sunlight is not very effective at directly warming the atmosphere. Most heating of the atmosphere occurs in the lowermost centimeter of the atmosphere, where air is in direct contact with Earth's surface. *Conduction* is the transfer of energy from one object to another by physical contact. As Earth's land and water surface is warmed by solar radiation, it imparts some of that energy to air molecules that collide with the surface. These air molecules gain kinetic energy, thus increasing the air temperature.

Once air is heated by conduction at Earth's surface, it begins to rise up in the atmosphere. This moving air carries heat from one place to another. *Convection* is the transfer of energy by the flow of heated fluids such as air. As warm air rises from the surface, cooler air flows in to replace it and it too is heated by the surface. The circulation of air associated with sea breezes, illustrated in Figure 13.11, is an example of convection. Over time, large quantities of air are warmed through the processes of radiation, conduction, and convection.

As shown in Figure 13.17, about 30% of the radiation that reaches Earth from the sun is reflected back into space by the atmosphere, clouds, or surface. This reflected radiation does nothing to heat the Earth or its atmosphere. An additional 19%—mostly infrared radiation—is absorbed by the atmosphere and clouds, which causes the atmosphere to become warmer. Much of the direct solar heating of the atmosphere occurs in the stratosphere. The remaining 51% of solar radiation is absorbed by Earth's surface. But Earth's surface doesn't just get warmer and warmer; all this surface heat is transferred to the atmosphere by conduction and radiation.

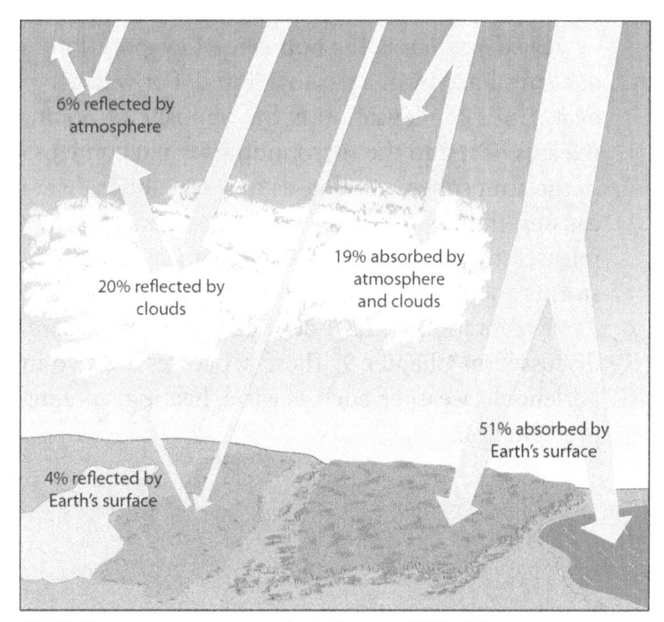

13.17. On average over a period of a year, 30% of the energy from the sun is reflected back into space, 19% is absorbed by the atmosphere and clouds, and 51% is absorbed by Earth's surface.

The average overall temperature of Earth's atmosphere does not change much over time. When it is winter in the northern hemisphere, it is summer in the southern hemisphere, and when it is day in the western hemisphere, it is night in the eastern hemisphere. In order for the average overall temperature of Earth to remain constant, the amount of energy that reaches Earth must be balanced by the amount of energy that leaves Earth.[7] As Earth gains energy from the sun, it continually loses the same amount of energy to space, almost entirely as invisible infrared radiation. If this balance did not occur, Earth would either get much hotter over time or much colder. In Chapter 15, we will address the topic of climate change and global warming. Any long-term warming that Earth's atmosphere does experience is very small compared to the overall amount of energy gained and lost by Earth every day.

13.3.2 Evaporation, Condensation, and Energy

You know that it takes energy to boil water. If you put water in a pot and put the pot on a stove, the water does not come to a boil unless you turn on the stove. But liquid water does not have to come to a boil in order for it to change its phase from liquid to gas. *Evaporation* is the phase change of a substance—in this case water—from liquid to gas without first being heated to the boiling point. In any body of water, whether a mud puddle or the ocean, some molecules have enough kinetic energy to break free from the water surface and enter the atmosphere. If the temperature of the water increases, more molecules have enough kinetic energy to escape from the liquid phase. Just as it takes heat to boil water, so also it takes heat to cause water to evaporate. On Earth's surface, almost all this heat comes from the sun. Evaporation occurs more rapidly on a warm day than on a cold day, and more rapidly on a sunny day than on a cloudy day. Most evaporation on Earth's surface comes from the oceans, so the oceans are the primary source for the water that goes into clouds and eventually falls on Earth as precipitation. Water also evaporates from lakes, streams, and soil. When plants lose water vapor to the atmosphere during photosynthesis, the process is called transpiration instead of evaporation.

Condensation is the opposite of evaporation or boiling; it is the phase change of a substance from a gas to a liquid. For water to boil or evaporate, it must absorb heat. Since condensation is the opposite of boiling and evaporation, water vapor releases heat into the surroundings when condensation occurs. When water vapor in the atmosphere condenses to form clouds, heat is released into the atmosphere, causing the temperature of the air to increase. The tremendous amount of heat released by condensation is the primary driver of severe storms such as thunderstorms and hurricanes.

Evaporation and condensation are both essential parts of the hydrologic cycle discussed in Chapter 9. These processes are also involved in much of what we experience as weather, such as wind, heating, cooling, cloud formation, fog, dew, and precipitation.

---

7 That is, not including the extremely small percentage (much less than 0.1%) of incident sunlight that is converted to different forms of energy by processes such as photosynthesis in plants and generation of electricity by solar cells.

Learning Check 13.3

1. Explain how radiation, conduction, and convection are all involved in the warming of the atmosphere on a clear, sunny day.
2. Compare and contrast boiling, evaporation, and condensation.

13.4 Circulation of the Atmosphere

Wind may blow from any direction, but at any given place, there are directions from which the wind most commonly blows. For instance, storms in the United States (outside of Alaska and Hawaii) generally move from west to east as they are carried along by larger wind systems.

13.4.1 Global Wind Systems

A much greater amount of solar radiation strikes Earth at the Equator than at the poles and this causes greater heating of Earth's surface in tropical regions. Just like other substances, air expands when it is heated. When air expands, it becomes less dense and pockets of air begin to ascend higher in the atmosphere. The result of these factors is that the area around the Equator is an area characterized by rising air.

Rising air cools down and sinking air warms up. The air that rises at the Equator is very cold by the time it reaches the top of the troposphere and it begins to move horizontally towards the poles. If Earth didn't rotate, this cold air from the tropics would probably blow as an upper-troposphere wind all the way to the poles, where it would sink. Polar air, on the other hand, would blow along the surface all the way to the Equator, warming as it goes. However, the Coriolis effect causes wind currents to curve to the right in the Northern Hemisphere and to the left in the Southern Hemisphere, as illustrated in Figure 13.18. The Coriolis effect also causes ocean currents to curve left or right, as described in Section

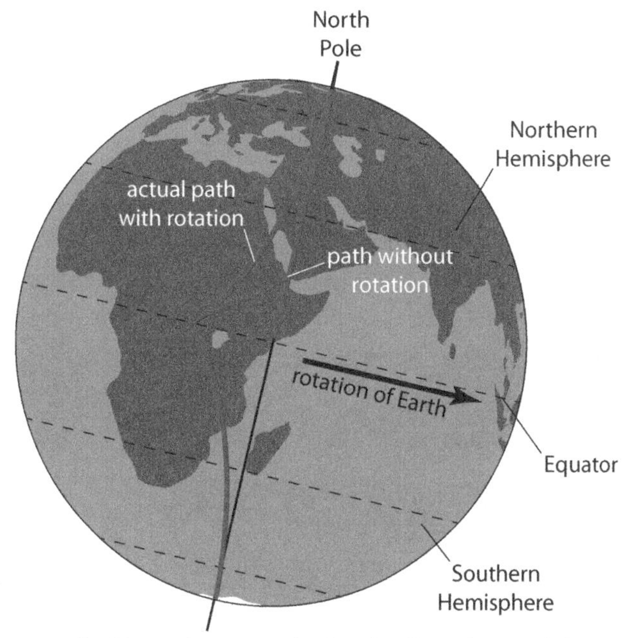

13.18. Earth's rotation causes the Coriolis effect. Along with the uneven heating of Earth's surface, this is part of the reason that Northern and Southern Hemispheres each have three global wind systems.

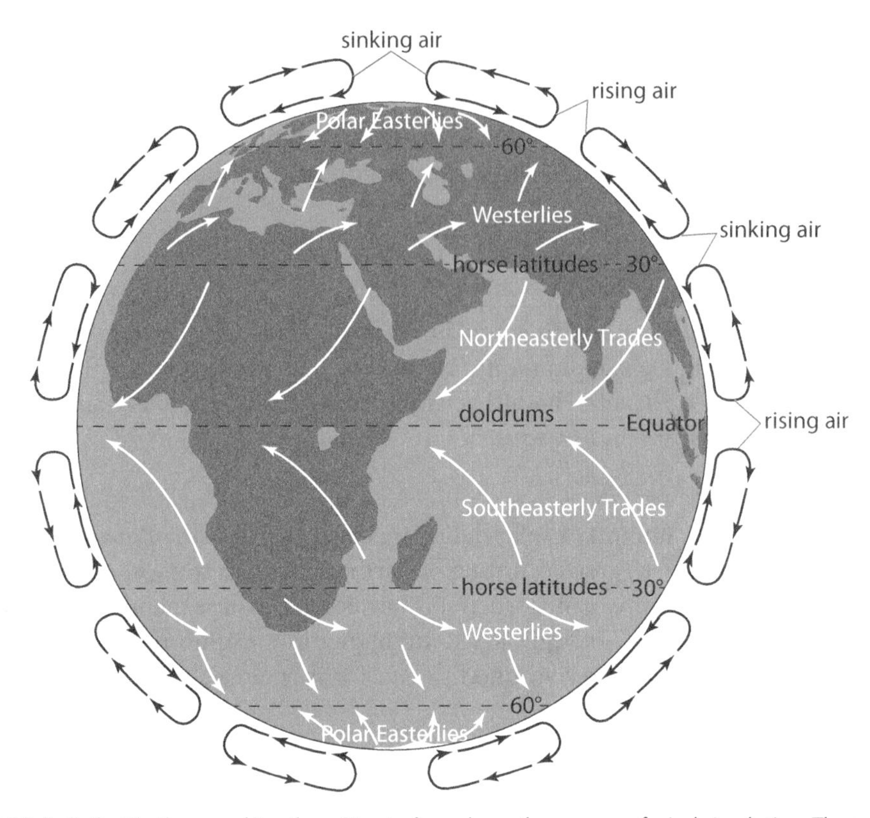

13.19. Both the Northern and Southern Hemispheres have three zones of wind circulation. The arrows indicating prevailing wind directions are curved because of the Coriolis effect.

12.3. The combined effect of the uneven heating of Earth's surface and the Coriolis effect is illustrated in Figure 13.19, which shows the global wind systems of Earth's atmosphere. There are three basic zones of air circulation in each hemisphere, divided at the Equator (0° latitude), 30° latitude, and 60° latitude.

Rising air at the Equator creates a band of cloudiness and precipitation, visible in the satellite image in Figure 13.20. This is the reason why many of Earth's tropical rain forests are at or near the Equator. This area is also characterized by generally weak winds, which makes sailing difficult. Mariners in the age of sail called this region the *doldrums*[8] because in this area sailing vessels could become nearly motionless for days at a time.

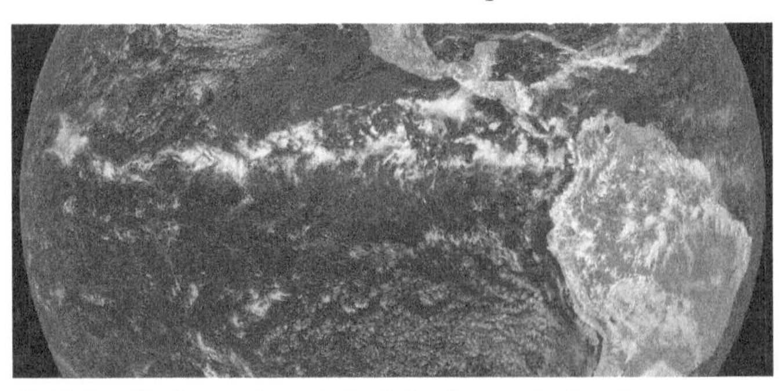

13.20. Satellite image shown belt of clouds near the Equator.

Sinking air at 30° north and south latitude

8 A more technical term for the doldrums is the intertropical convergence zone (ITCZ).

becomes warm and dry as it sinks, creating extensive bands of deserts around the globe, such as the Sahara and Arabian Deserts in the northern hemisphere, and the Kalahari Desert in southern Africa. These areas of sinking air are also areas of relatively calm winds, which cause difficulties for sailing vessels. Mariners named this region of the seas the *horse latitudes*. (One widely-told story—probably not true—for the origin of this name is that Spanish trading ships became becalmed in these waters, and horses onboard died of thirst and had to be thrown overboard.) The areas between the Equator and 30° latitude are known for the *trade winds*. In the Northern Hemisphere, the trade winds generally blow from the northeast; in the Southern Hemisphere, they generally blow from the southeast, as indicated by the arrows in Figure 13.19.

The second wind system, known as the *prevailing westerlies*, lies between 30° and 60° latitude. The wind patterns in the prevailing westerlies are opposite to that of the trade winds, with average wind direction in the Northern Hemisphere being from the southwest. The prevailing westerlies are responsible for most of the weather systems that move across mid-latitude areas such as the United States (outside of Hawaii and most of Alaska) and Europe.

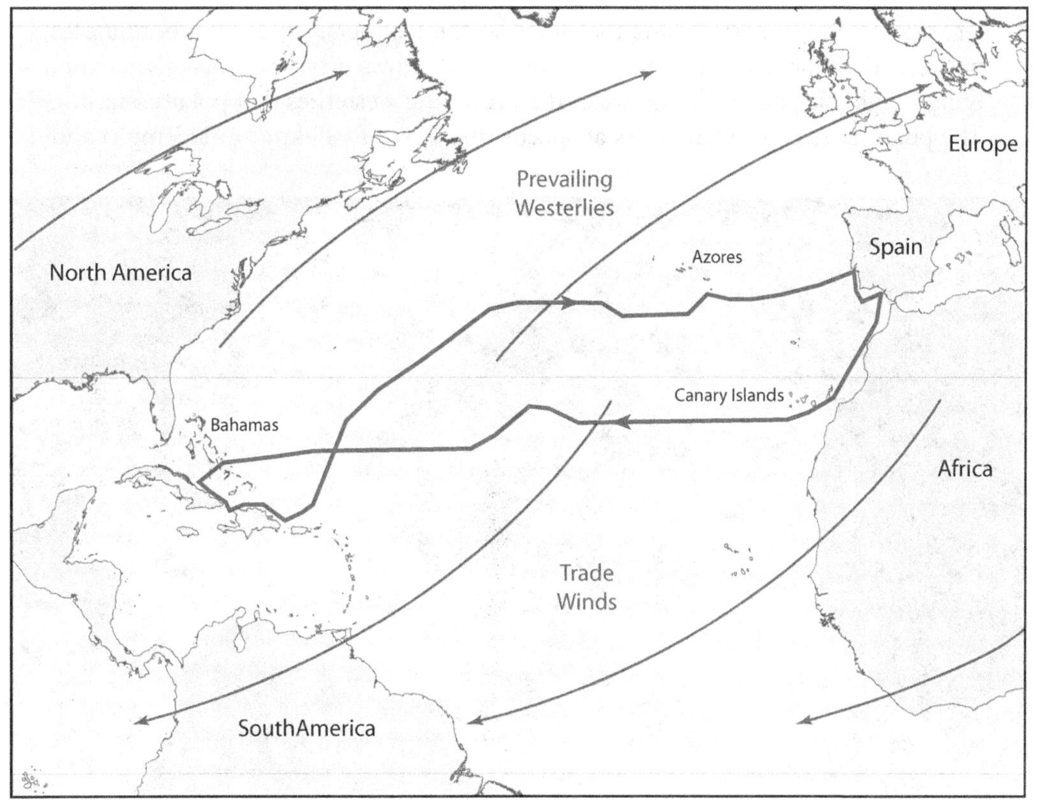

13.21. The first voyage of Christopher Columbus, 1492–1493. The ships sailed in the trade winds, which are easterlies, on the trip to the New World, and in the prevailing westerlies for most of the return trip.

The third wind system, the *polar easterlies*, blows between 60° latitude and the poles. In the Northern Hemisphere, the polar easterlies blow from the northeast. The boundary between the prevailing westerlies and polar easterlies is called the *polar front*. Unlike the doldrums and horse latitudes, weather along the polar front is often stormy. This is due to a sharp difference in temperature between air on the two sides of the polar front. During winter, the polar front moves somewhat further away from the poles, bringing surges of bitterly cold air into mid-latitude regions.

In the age of sail, explorers and traders used knowledge of these wind systems to power their ships across oceans. Figure 13.21 shows the path taken by Christopher Columbus on his first voyage, a route typical of the course taken by later trading ships traveling between Europe and North America. Ships leaving Europe sailed south along the coast of North Africa until they reached the trade winds, which they used to propel themselves westward across the Atlantic Ocean. On the return trip, they took a more northerly route to take advantage of the prevailing westerlies. It wasn't until the 1850s, near the end of the age of sail, that Matthew Maury published the first global maps of the three wind systems. As you recall from Chapter 12, Maury was also the first to chart worldwide ocean currents.

13.4.2 Jet Streams

A *jet stream* is a narrow, high-altitude band of strong winds. The most common jet streams are westerly winds that form at the tropopause near the boundaries between the global wind systems. Figure 13.22 shows a typical polar jet stream, lying roughly at the border between the prevailing westerlies and polar easterlies. The polar jet stream often blows at speeds in excess of 185 km/hr (115 mph) and

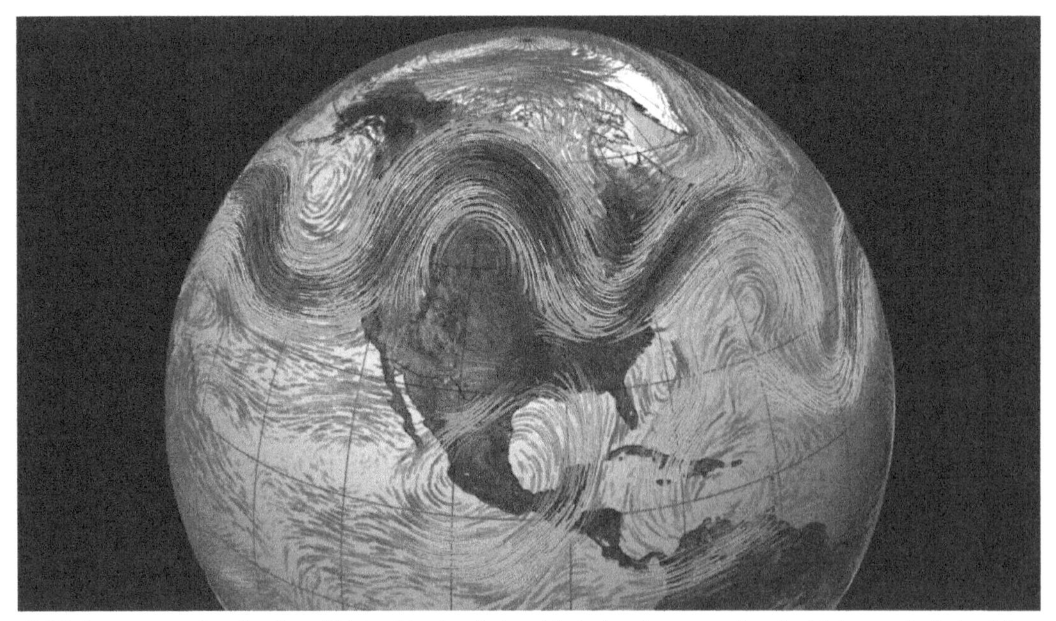

13.22. A computer visualization of high-altitude winds, with dark red representing the high-speed winds of the polar jet stream. The path of the jet stream continuously changes.

sometimes blows even more strongly. Jet streams are streams of air that may be thousands of kilometers long, but are generally only a few hundred kilometers wide and a few kilometers thick.

Jet airliners travelling long distances, such as across the Atlantic or Pacific Oceans, can cut an hour or more off their travel time and save tremendous amounts of fuel by either flying in the jet stream when flying west to east or avoiding the headwinds of the jet stream when flying east to west. Aircraft even fly hundreds of kilometers away from the shortest flight path in order to either catch or avoid a jet stream.

Learning Check 13.4

1. Describe the three main wind systems that exist in each of the Northern and Southern Hemispheres.
2. Explain why tropical rain forests are located near the Equator.
3. Explain why deserts are common near 30° north and south of the Equator.
4. Describe how the Coriolis effect affects wind in the Northern and Southern Hemispheres.
5. What is a jet stream and where are jet streams found?

Chapter 13 Exercises

Answer each of the questions below as completely as you can. Write your responses in complete sentences unless instructed otherwise.

1. What would Earth be like if solar radiation could not travel through the empty vacuum of outer space?
2. Speculate about what would have happened to Earth's ozone layer if use of CFCs had continued to increase rather than being sharply reduced by an international agreement. What would have been the consequences of allowing this to happen?
3. Compare and contrast how the troposphere and stratosphere are warmed.
4. Why do mountain climbers on Mt. Everest, the tallest mountain on Earth, need to wear oxygen masks?
5. Navigators on sailing ships take advantage of wind currents to cross the Atlantic Ocean; pilots of modern aircraft do the same. Compare and contrast the wind currents these travelers use to plan their voyages.

CPSIA information can be obtained
at www.ICGtesting.com
Printed in the USA
FSOW03n1944160116
15809FS